# STEM CELLS

**Dr. Logeswari Selvaraj**
Professor of Microbiology
Thanjavur Medical College
Thanjavur

**MJP Publishers**

Chennai New Delhi Tirunelveli

ISBN 978-81-8094-199-3

Printed and bound in India

**MJP Publishers**
New No. 5 Muthu Kalathy Street,
Triplicane,
Chennai 600 005

MJP 174


Publisher : J.C. Pillai

*To*

*My Parents Mrs. & Mr. Selvaraj*

*and*

*My Children Stephen and Simeon*

# Preface

The basic definition of stem cells is that they are the engine room of multicellular organisms for both plants and animals. These cells are surrounded by intricate signals that allow them to divide so as to self-renew and to produce a progeny to make a specific type of tissue. In 1855, Rudolph Virchow, a German pathologist, explained the idea that all living things come from other living cells and thus paved the way for the current definition of stem cells.

Initially it was described that the embryonic carcinoma cells are the primitive stem cells that can make all types of body tissue—the hair, bone and the brain. In 1896, E.D.Wilson was the first person who coined the term "stem cell". In 1950's the stem cell science was first initiated when bone marrow was transplanted into irradiated mice and shown to reconstitute the stem cell population. Thereafter several scientists showed research interests using stem cells in the molecular processes of neurological diseases and drug delivery systems *in vivo*.

In 2007, researchers showed the 'universal' type of stem cell that was isolated and characterized from mouse embryos. These embryonic stem cells divided in *in vitro* cultures maintaining the potential to create the tissues in one system. This research on stem cells provided a widespread data which enlightens to learn every basic concepts about stem cells.

This book contains the basics of stem cells, their scientific aspects in the latest applied areas. The chapters are presented in a sequential organization for effective learning. It is hopeful that this book provides detail for the needs of students and faculty members.

**Dr. Logeswari Selvaraj**

# Contents

# 1

# Introduction to Stem Cell

Stem cells are biological cells found in all multicellular organisms that can divide (through mitosis) and differentiate into diverse specialized cell types and can self-renew to produce more stem cells. In mammals, there are two broad types of stem cells: embryonic stem cells, which are isolated from the inner cell mass of blastocysts, and adult stem cells, which are found in various tissues. In adult organisms, stem cells and progenitor cells act as a repair system for the body, replenishing adult tissues. In a developing embryo, stem cells can differentiate into all the specialized cells, (these are called pluripotent cells) but also maintain the normal turnover of regenerative organs, such as blood, skin, or intestinal tissues.

All stem cells are unspecialized (undifferentiated) cells that are characteristically of the same family type (lineage). They retain the ability to divide throughout life and give rise to cells that can become highly specialized and take the place of cells that die or are lost. Stem cells contribute to the body's ability to renew and repair its tissues. Unlike mature cells, which are permanently committed to their fate, stem cells can both renew themselves as well as create new cells of whatever tissue they belong to (and other tissues). Bone marrow stem cells, for example, are the most primitive cells in the marrow. From them all the various types of blood cells are descended. Bone marrow stem-cell transfusions (or transplants) were originally given to replace various types of blood cells. Stem cells can now be artificially grown and transformed into specialized cell types with characteristics consistent with cells of various tissues such as muscles or nerves through cell culture. Highly plastic adult stem cells are routinely used in medical therapies. Stem cells can be taken from a variety of sources, including umbilical cord blood and bone marrow. Embryonic cell lines and autologous embryonic stem cells generated through therapeutic cloning have also been proposed as promising candidates for future therapies. There are three sources of autologous adult stem cells. 1) Bone marrow, which requires extraction by harvesting, that is, drilling into bone (typically the femur of iliac crest), 2) Adipose tissue (lipid cells), which requires extraction by liposuction, and 3) Blood, which requires extraction through pheresis, wherein blood is drawn from the donor, (similar to a blood donation) passed through a machine that extracts the stem cells and returns other portions of the blood to the donor. Of all stem cell types, autologous harvesting involves the least risk. By definition, autologous cells are obtained from one's own body, just as one may bank his or her own blood for elective surgical procedures.

## PROPERTIES

The classical definition of a stem cell requires that it has two properties:

- **Self-renewal** The ability to go through numerous cycles of cell division while maintaining the undifferentiated state. Two mechanisms exist to ensure that the stem cell population is maintained:

  1. **Obligatory asymmetric replication** A stem cell divides into one father cell that is identical to the original stem cell, and another daughter cell that is differentiated.

  2. **Stochastic differentiation** When one stem cell develops into two differentiated daughter cells, another stem cell undergoes mitosis and produces two stem cells identical to the original.

- **Potency** The capacity to differentiate into specialized cell types. In the strictest sense, this requires stem cells to be either totipotent or pluripotent (to be able to give rise to any mature cell type), although multipotent or unipotent progenitor cells are sometimes referred to as stem cells.

## IDENTIFICATION

The practical definition of a stem cell is the functional definition—a cell that has the potential to regenerate tissue over a lifetime. For example, the defining test for a bone marrow or hematopoietic stem cell (HSC) is the ability to transplant one cell and save an individual without HSCs. In this case, a stem cell must be able to produce new blood cells and immune cells over a long term, demonstrating potency. It should also be possible to isolate stem cells from the transplanted individual, which can themselves be transplanted into another individual without HSCs, demonstrating that the stem cell was able to self-renew. Properties of stem cells can be illustrated *in vitro*, using methods such as clonogenic assays, in which single cells are assessed for their ability to differentiate and self-renew. Stem cells can also be isolated by their possession of a distinctive set of cell surface markers.

## TYPES OF STEM CELLS

Stem cells can be classified into various types, based on their origin from where they are obtained.

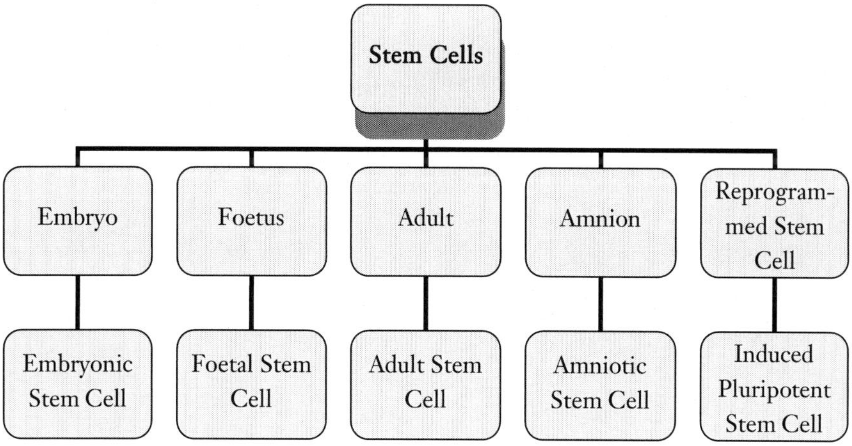

**Embryonic** Embryonic stem cell lines (ES cell lines) are cultures of cells derived from the epiblast tissue of the inner cell mass (ICM) of a blastocyst or earlier morula stage embryos. A blastocyst is an early stage embryo approximately four to five days old in humans and consisting of 50–150 cells. ES cells are pluripotent and give rise during development to all derivatives of the three primary germ layers: ectoderm, endoderm and mesoderm. In other words, they can develop into each of the more than 200 cell types of the adult body when given sufficient and necessary stimulation for a specific cell type. They do not contribute to the extra-embryonic membranes or the placenta. The endoderm is composed of the entire gut tube and the lungs, the ectoderm gives rise to the nervous system and skin, and the mesoderm gives rise to muscle, bone, and blood. Nearly all research to date has made use of mouse embryonic stem cells (mES) or human embryonic stem cells (hES). Both have the essential stem cell characteristics. Yet they require very different environments in order to maintain an undifferentiated state. Mouse ES cells are grown on a layer of gelatin and require the presence of leukaemia inhibitory factor (LIF). Human ES cells are grown on a feeder layer of mouse embryonic fibroblasts (MEFs) and require the presence of basic fibroblast growth factor (bFGF or FGF-2). A human embryonic stem cell is also defined by the presence of several transcription factors and cell surface proteins. The transcription factors Oct-4, Nanog and Sox2 form the core regulatory network that ensures the suppression of genes that lead to differentiation and the maintenance of pluripotency. The cell surface antigens most commonly used to identify hES cells are the glycolipids SSEA3 and SSEA4 and the keratan sulphate antigens Tra-1-60 and Tra-1-81. The molecular definition of a stem cell includes many more proteins and continues to be a topic of research. There are currently no approved treatments using embryonic stem cells. The first human trial was approved by the US Food and Drug Administration in January 2009. However, the human trial had not yet been initiated until October 13, 2010 in Atlanta for spinal injury victims. ES cells, being pluripotent cells, require specific signals for correct differentiation if injected directly into another body. ES cells will differentiate into many different types of cells, causing a teratoma. Many nations currently have moratoria on either ES cell research or the production of new ES cell lines. Because of their combined abilities of unlimited expansion and pluripotency, embryonic stem cells remain a theoretically potential source for regenerative medicine and tissue replacement after injury or disease.

**Foetal** The primitive stem cells located in the organs of foetuses are referred to as foetal stem cells.

**Adult** Adult Stem Cells are also known as somatic (from the Greek word "somatikos" meaning "of the body") stem cells and germ line (giving rise to gametes) stem cells, they can be found in children, as well as adults.

Pluripotent adult stem cells are rare and generally small in number but can be found in a number of tissues including umbilical cord blood. A great deal of adult stem cell research to date has had the aim of characterizing the capacity of the cells to divide or self-renew indefinitely and their differentiation potential. In mice, pluripotent stem cells are directly generated from adult fibroblast cultures. Most adult stem cells are lineage-restricted (multipotent) and are generally referred to by their tissue origin (mesenchymal stem cell, adipose-derived stem cell, endothelial stem cell, dental pulp stem cell, etc.). Adult stem cell treatments have been successfully used for many years to treat leukaemia and related bone/blood cancers through bone marrow transplants. Adult stem cells are also used in veterinary medicine to treat tendon and ligament injuries in horses. The use of adult stem cells in research and therapy is not as controversial as the use of embryonic stem cells, because the production of adult stem cells does not require the destruction of an embryo. Additionally, in instances where adult stem cells are obtained from the intended recipient (an autograft), the risk of rejection is essentially non-existent. An extremely rich source for adult mesenchymal stem cells is the developing tooth bud of the mandibular third molar. The stem cells eventually form enamel (ectoderm), dentin, periodontal ligament, blood vessels, dental pulp, nervous tissues, and a minimum of 29 different end organs. Because of extreme ease in collection at 8–10 years of age before calcification and minimal to no morbidity, these will probably constitute a major source of cells for personal banking, research and current or future therapies. These stem cells have been shown capable of producing hepatocytes.

**Amniotic** Multipotent stem cells are also found in amniotic fluid. These stem cells are very active, expand extensively without feeders and are not tumorigenic. Amniotic stem cells are multipotent and can differentiate in cells of adipogenic, osteogenic, myogenic, endothelial, hepatic and also neuronal lines.

Use of stem cells from amniotic fluid overcomes the ethical objections to using human embryos as a source of cells. It is possible to collect amniotic stem cells for donors or for autologous use: the first US amniotic stem cells bank was opened in 2009 in Medford, MA by Biocell Center Corporation and collaborates with various hospitals and universities all over the world.

## Induced Pluripotent

These are not adult stem cells, but rather reprogrammed cells (e.g., epithelial cells) given pluripotent capabilities. Using genetic reprogramming with protein transcription factors, pluripotent stem cells equivalent to embryonic stem cells have been derived from human adult skin tissue. Shinya Yamanaka and his colleagues at Kyoto University used the transcription factors Oct3/4, Sox2, c-Myc, and Klf4 in their experiments on cells from human faces. As a result of the success of these experiments, Ian Wilmut, who helped create the first cloned animal Dolly the Sheep, has announced that he will abandon nuclear transfer as an avenue of research. Frozen blood samples can be used as a source of induced pluripotent stem cells, opening a new avenue for obtaining the valued cells.

## Lineage

To ensure self-renewal, stem cells undergo two types of cell division. Symmetric division gives rise to two identical daughter cells both endowed with stem cell properties. Asymmetric division, on the other hand, produces only one stem cell and a progenitor cell with limited self-renewal potential. Progenitors can go through several rounds of cell division before terminally differentiating into a mature cell. It is possible that the molecular distinction between symmetric and asymmetric divisions lies in differential segregation of cell membrane proteins (such as receptors) between the daughter cells. An alternative theory is that stem cells remain undifferentiated due to environmental cues in their particular niche. Stem cells differentiate when they leave that niche or no longer receive those signals. Studies in Drosophila germarium have identified the signals dpp and adherens junctions that prevent germarium stem cells from differentiating. The signals that lead to reprogramming of cells to an embryonic-like state are also being investigated. These signal pathways include several transcription factors including the oncogene c-Myc. Initial studies indicate that transformation of mice cells with a combination of these anti-differentiation signals can reverse differentiation and may allow adult cells to become pluripotent. It was recently determined that the somatic expression of combined transcription factors can directly induce other defined somatic cell fates; researchers identified three neural-lineage-specific transcription factors that could directly convert mouse fibroblasts (skin cells) into fully functional neurons. This "induced neurons" (iN) cell research inspires the researchers to induce other cell types. It implies that all cells are totipotent: with the proper tools, all cells may form all kinds of tissue.

## USE OF STEM CELLS IN THERAPHY

Medical researchers believe that stem cell therapy has the potential to dramatically change the treatment of human disease. A number of adult stem cell therapies already exist, particularly bone marrow transplants that are used to treat leukaemia. Technologies derived from stem cell research is able to treat a wider variety of diseases including cancer, Parkinson's disease, spinal cord injuries, Amyotrophic lateral sclerosis, multiple sclerosis and muscle damage, amongst a number of other impairments and conditions. Stem cells are widely studied, for their potential therapeutic use and for their inherent interest. It has been proposed that surplus embryos created for *in vitro* fertilization could be donated with consent and used for the research.

## SOME FACTS ABOUT STEMCELLS

1. What are embryonic stem cells?

   Embryonic stem (ES) cells are a collection of cells found only in very early development which are the precursors to every cell type in the human body. The vast majority of cells in the body (somatic cells) fall into specific classes or types, such as muscle, bone and neurons, each of which have unique characteristics and functions. However, these cells are not interchangeable (a muscle cell cannot become a neuron) and most of these cells have lost the ability to multiply to create new cells. ES cells differ from all other cells in two important ways. First, they can be induced to change, or differentiate, into virtually any cell type. Second, unlike somatic cells which have finite life spans, ES cell can grow indefinitely in culture. These two unique characteristics give ES cells enormous potential for application in medicine and science.

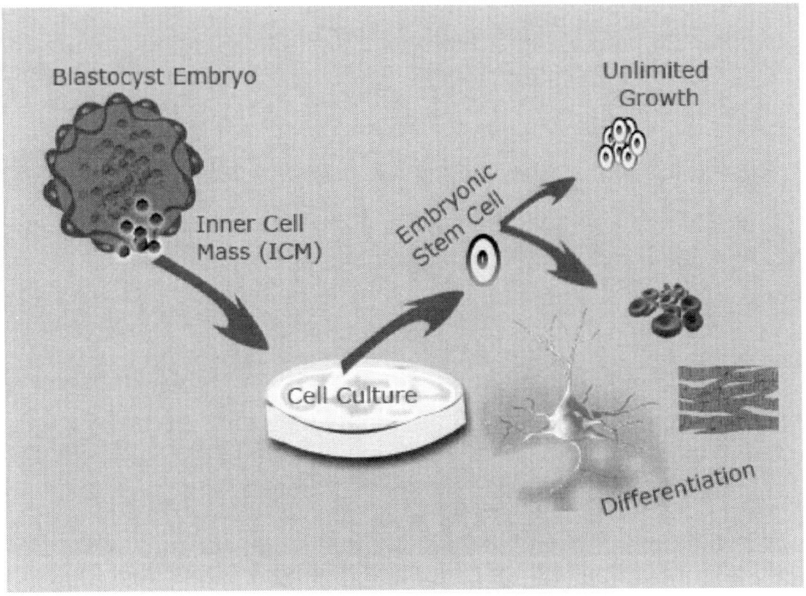


2. Why are embryonic stem cells so important?

Embryonic stem cells are important to medicine because of their ability to change into other cell types. This ability means that ES cells have the potential to repair damaged organs and replace cells that do not function properly. Since they can multiply indefinitely, the large numbers of cells necessary to repair or replace these tissues can be produced. Thus, the hope is that ES cells can be a renewable source of replacement cells that can be used to treat a number of medical problems including Parkinson's and Alzheimer's disease, strokes, burns, spinal cord damage and heart disease.

3. What are iPS cells?

Recent publications have described the derivation of ES-like induced pluripotent stem (iPS) cells from adult mouse and human cells (Nakagawa *et al.* 2008, Takahashi *et al.* 2007; Yu *et al*, 2007). These researchers introduced specific sets of genes encoding transcription factors which are normally expressed in undifferentiated ES cells. The expression of these genes resulted in the "reprogramming" of the adult cells to a more ES-like or pluripotent state. In addition, the reintroduction of these genes can have adverse consequences. For instance, the use of retroviruses and the potential for reactivation of introduced genes such as c-myc and Oct-4 can increase the risk of cancer. These issues will need to be addressed if iPS technology will have clinical application.

4. What are adult stem cells?

The human body has a relatively small number of cells, called adult stem cells that are capable of differentiating into a limited range of cell types. For instance, blood stem cells are capable of changing into a number or different types of blood cells. These adult ES cells are also of enormous importance to medicine. However, they have limitations that ES cells do not have. First, they are limited in the number of types of cells into which they can change. For instance, blood stem cells cannot form bone. In addition, unlike ES cells adult stem cells do not appear to have the same capacity to multiply indefinitely. They have also been more difficult to grow in the laboratory. So, while adult stem cells are important, they cannot be viewed at this time as a replacement for ES cells. Research into all types of stem cells is needed in order to develop medicine's ability to treat disease.

5. Why do we have stem cells in our body?

Each cell type has a specific job in our body: for example, some brain cells (neurons) make sure that information is processed and transmitted properly so that we are able to learn, reason, and speak. Our red blood cells (erythrocytes) help carry oxygen throughout our body. Because stem cells can differentiate into any mature cell types, they can replace worn out, damaged or diseased cells and promote the growth of our tissues.

6. Are there different types of stem cells?

   To date, scientists have discovered at least four types of stem cells, and there may be even more in the future. The four current types of stem cells are categorized and named after their source: Embryonic Stem Cells (ESCs), Adult Stem Cells, Umbilical Cord Stem Cells, and Induced Pluripotent Stem Cells (iPSCs).

## BASICS OF STEM CELL THERAPY

It is an innovative area of medical research in Regenerative Medicine with greater potential to benefit the treatment of many incurable diseases and damaged organs.

## SIGNIFICANCE OF STEM CELLS IN THERAPEUTICS

Stem cells are non-specialized cells that have the capacity to divide in culture for longer period of time and also differentiate into more matured cells with specialized functions and with self-renewal capacity.

## BASICS, STEM CELLS, SCIENCE AND SOCIETY

Stem cells are full of potential. They can become blood cells, heart cells, skin cells, brain cells and bone cells, in fact they can develop into many or all over 200 different types of specialized cells that make up our body. Stem cells have two important characteristics that set them apart from our other body cells. They are able to divide and produce more stem cells (self-renew) over long periods of time or indefinitely (an immortal cell line). On receiving the right signals, stem cells can be stimulated to become specialized cells (this process is called differentiation).

## TYPES OF STEM CELLS

Stem cells can be generally classified into three types.

1. **Embryonic stem cells** Embryonic stem cells are derived from embryos that develop from eggs that have been fertilized through IVF. The embryos from which embryonic stem cells are derived are typically four or five days old and are a hollow microscopic ball of cells called the blastocyst. It is the richest source of undifferentiated stem cells.

2. **Adult stem cells** Adult stem cells are partially undifferentiated cells found in a tissue in an adult organism that can renew themselves and may, within certain limitations, differentiate to yield all of the specialized cell types of the tissue from which they originated.

3. **Umbilical cord blood stem cells** Stem cells that are abundantly found in the new born child's cord blood are known as Umbilical Cord Stem Cells. They hold great potential for medical application in future.

7. Can stem cells be used as a therapy?

Ultimately, it is hoped that stem cells may be used to replace cells or regenerate organs that are diseased or failing in our body, such as replacing destroyed heart tissue in people with heart disease, or regenerating brain cells that are lost due to old age, substance abuse, or neurodegenerative diseases.

8. Which type of stem cells do we normally use?

We normally use autologous adult stem cells where there is no possibility of rejection by the immune system of the human body as the cells are drawn from the patient's own body.

9. Where are adult stem cells found?

The best known source in the body for adult stem cells is in bone marrow. It is also found in blood, eye, brain, skeleton muscles, liver, pancreas, lumen and skin.

10. What is the function of stem cells?

    Irrespective of the origin of stem cells, they have the ability to continuously divide and differentiate into various other kinds of cells to repair and regenerate damaged cells and wherever the body tissues are damaged, the stem cells automatically head for such areas for repairing or replacing the damaged cells.

11. Why do we not see repair in some diseases?

    If the disease process is too rapid and massive or if the stem cells are not able to reach the target organ in sufficient quantity, the treatment by stem cells therapy is recommended.

# 2

# Embryonic Stem Cells

Embryonic stem cells (ES cells), as their name suggests, are derived from embryos. Most embryonic stem cells are derived from embryos that develop from eggs that have been fertilized *in vitro*—in an IN VITRO fertilization clinic—and then donated for research purpose with informed consent of the donors. They are not derived from fertilized eggs in the womb of the mother. Embryonic stem cells (ES cells) are pluripotent stem cells which means embryonic stem cells are able to develop into any type of cell, excepting those of the placenta. Only embryonic stem cells of the morula are totipotent which are able to develop into any type of cell, including those of the placenta. ES cells are derived from the inner cell mass of the blastocyst of an early-stage embryo. Human embryos reach the blastocyst stage by 4–5 days post fertilization, at which time they consist of 50–150 cells by isolating the embryoblast cell mass.

## RESEARCH HISTORY AND DEVELOPMENTS

In 1964, researchers isolated a single type of cell from a teratocarcinoma, a tumour now known to be derived from a germ cell. These cells isolated from the teratocarcinoma replicated and grew in cell culture as a stem cell and are now known as embryonic carcinoma (EC) cells. In 1981, embryonic stem cells (ES cells) were independently first derived from mouse embryos by two groups. Martin Evans and Matthew Kaufman from the Department of Genetics, University of Cambridge published first in July, revealing a new technique for culturing the mouse embryos in the uterus to allow for an increase in cell number, allowing for the derivation of ES cells from these embryos.

## CHARACTERISTICS OF ES CELLS

Embryonic stem cells are distinguished by two distinctive properties:

- Pluripotency, and
- Ability to replicate indefinitely.

ES cells are pluripotent, that is, they are able to differentiate into all derivatives of the three primary germ layers: ectoderm, endoderm, and mesoderm. These include each of the more than 220 cell types in the adult body. Pluripotency distinguishes embryonic stem cells from adult stem cells found in adults; while embryonic stem cells can generate all cell types in the body, adult stem cells are multipotent and can only produce a limited number of cell types. Additionally, under defined conditions, embryonic stem cells are capable of propagating themselves indefinitely. This allows embryonic stem cells to be employed as useful tools for both research and regenerative medicine, because they can produce limitless numbers of themselves for continued research or clinical use. ES cell therapies have been proposed for regenerative medicine and tissue replacement after injury or disease due to their property of plasticity and potentially unlimited capacity for self-renewal. Diseases that could potentially be treated by pluripotent stem cells include a number of blood and immune-system related genetic diseases, cancers and disorders; juvenile diabetes; Parkinson's disease; blindness and spinal cord injuries.

## CONCEPT OF CELL CULTURE

Growing cells in the laboratory is known as cell culture. Human embryonic stem cells (hESCs) are generated by transferring cells from a preimplantation-stage embryo into a plastic laboratory culture dish that contains a nutrient broth known as culture medium. The cells divide and spread over the surface of the dish. The inner surface of the culture dish is typically coated with mouse embryonic skin cells that have been treated so they will not divide. This layer coated with mouse embryonic skin cells is called a feeder layer. The mouse cells in the bottom of the culture dish provide the cells a sticky surface to which they can attach. Also, the feeder cells release nutrients into the culture medium. Researchers have devised ways to grow embryonic stem cells without mouse feeder cells. However, if the plated cells survive, divide and multiply enough to crowd the dish, they are removed gently and plated into several fresh culture dishes. The process of re-plating or subculturing the cells is repeated many times and for many months. Each cycle of subculturing the cells is referred to as a passage. Once the cell line is established, the original cells yield millions of embryonic stem cells. Embryonic stem cells that have proliferated in cell culture for a prolonged period of time without differentiating, are pluripotent, and those that have not developed genetic abnormalities are referred to as an embryonic stem cell line. At any stage in the process, batches of cells can be frozen and shipped to other laboratories for further culture and experimentation.

## TECHNIQUES AND CONDITIONS FOR EMBRYONIC STEM CELL DERIVATION AND CULTURE

Embryonic stem cells are derived from the inner cell mass of the early embryo, which are harvested from the donor mother animal. Martin Evans and Matthew Kaufman reported a technique that delays embryo implantation, allowing the inner cell mass to increase. This process includes removing the donor mother's ovaries and dosing her with progesterone, changing the hormone environment, which causes the embryos to remain free in the uterus. After 4–6 days of this intrauterine culture, the embryos are harvested and grown in *in vitro* culture until the inner cell mass forms "egg cylinder-like structures," which are dissociated into single cells, and plated on fibroblasts treated with mitomycin-c (to prevent fibroblast mitosis). Clonal cell lines are created by growing up a single cell. Evans and Kaufman showed that the cells grown out from these cultures could form teratomas and embryoid bodies, and differentiate *in vitro*, which all indicate the cells are pluripotent. Gail Martin derived and cultured her ES cells differently. She removed the embryos from the donor mother at approximately 76 hours after copulation and cultured them overnight in media containing serum. The following day, she removed the inner cell mass from the late blastocyst using microsurgery. The extracted inner cell mass was cultured on fibroblasts treated with mitomycin-c in media containing serum and was conditioned by EC cells. After approximately one week, colonies of cells grew out. These cells grew in culture and demonstrated pluripotent characteristics, as demonstrated by the ability to form teratomas, differentiate *in vitro*, and form embryoid bodies. Martin referred to these cells as ES cells. It is now known that the feeder cells provide leukemic inhibitory factor (LIF) and serum provides bone morphogenetic proteins (BMPs) that are necessary to prevent ES cells from differentiating. These factors are extremely important for the efficiency of deriving ES cells. Further more, it has been demonstrated that different mouse strains have different efficiencies for isolating ES cells. Current uses for mouse ES cells include the generation of transgenic mice including knock out mice. For human treatment, there is a need for patient specific pluripotent cells.

## LABORATORY TO IDENTIFY EMBRYONIC STEM CELLS

During the process of generating embryonic stem cell lines, scientists test the cells to see whether they exhibit the fundamental properties that make them embryonic stem cells. This process is called characterization. There are no specific test to measure the fundamental properties and nature of ES. However, laboratories that grow human embryonic stem cell lines use several kinds of tests.

1. Growing and subculturing the stem cells for many months. This ensures that the cells are capable of long-term growth and self-renewal. Scientists inspect the cultures through a microscope to see that the cells look healthy and remain undifferentiated.

2. Using specific techniques to determine the presence of transcription factors that are typically produced by undifferentiated cells. Two of the most important transcription factors are Nanog and Oct4. Transcription factors help turn genes on and off at the right time, which is an important part of the processes of cell differentiation

and embryonic development. In this case, both Oct4 and Nanog are associated with maintaining the stem cells in an undifferentiated state and their capacity of self-renewal.

3. Using specific techniques to determine the presence of particular cell surface markers that are typically produced by undifferentiated cells.

4. Examining the chromosomes under a microscope. This is a method to assess whether the chromosomes are damaged or if the number of chromosomes has changed. It does not detect genetic mutations in the cells.

5. Determining whether the cells can be re-grown, or subcultured, after freezing, thawing, and re-plating.

6. Testing whether the human embryonic stem cells are pluripotent by 1) allowing the cells to differentiate spontaneously in cell culture; 2) manipulating the cells so they will differentiate to form cells characteristic of the three germ layers; or 3) injecting the cells into a mouse with a suppressed immune system to test for the formation of a benign tumor called a teratoma. Since the mouse's immune system is suppressed, the injected human stem cells are not rejected by the mouse immune system and scientists can observe growth and differentiation of the human stem cells.

## DIFFERENTIATION OF EMBRYONIC STEM CELLS

As long as the embryonic stem cells in culture are grown under appropriate conditions, they can remain undifferentiated (unspecialized). But if cells are allowed to clump together to form embryoid bodies, they begin to differentiate spontaneously. They can form muscle cells, nerve cells, and many other cell types. Although spontaneous differentiation is a good indication that a culture of embryonic stem cells is healthy, it is not an efficient way to produce cultures of specific cell types. To generate cultures of specific types of differentiated cells-heart muscle cells, blood cells, or nerve cells several methods are used. For example, the chemical composition of the culture medium is changed, the surface of the culture dish is altered, or the cell is modified by inserting specific genes. Through years of experimentation, scientists have established some basic protocols or "recipes" for the directed differentiation of embryonic stem cells into some specific cell types. If scientists can reliably direct the differentiation of embryonic stem cells into specific cell types, they may be able to use the resulting, differentiated cells to treat certain diseases in the future. Diseases that might be treated by transplanting cells generated from human embryonic stem cells include Parkinson's disease, diabetes, traumatic spinal cord injury, Duchenne's muscular dystrophy, heart disease, and vision and hearing loss.

# 3

# Adult Stem Cells

Adult stem cells are undifferentiated cells, found throughout the body after embryonic development. These cells multiply by cell division to replenish dying cells and regenerate damaged tissues. These are known as somatic stem cells, they can be found in juvenile as well as adult animals and humans. Scientific interest in adult stem cells is centered on their ability to divide or self-renew indefinitely, and generate all the cell types of the organ from which they originate, potentially regenerating the entire organ from a few cells. Unlike embryonic stem cells, the use of adult stem cells in research and therapy is not considered to be controversial, as they are derived from adult tissue samples rather than destroyed human embryos. They have mainly been studied in humans and model organisms such as mice and rats.

## PROPERTIES OF ADULT STEM CELLS

A stem cell possesses two properties:

- Self-renewal, which is the ability to go through numerous cycles of cell division while still maintaining its undifferentiated state.
- Multipotency or multidifferentiative potential, is the ability to generate progeny of several distinct cell types, (for example, both glial cells and neurons) as opposed to unipotency, which is the

term for cells that are restricted to producing a single-cell type. These properties can be illustrated with relative ease *in vitro*, using methods such as clonogenic assays, where the progeny of a single cell is characterized.

## Lineage

To ensure self-renewal, stem cells undergo two types of cell division. Symmetric division gives rise to two identical daughter cells, both endowed with stem cell properties, whereas asymmetric division produces only one stem cell and progenitor cell with limited self-renewal potential. Progenitors can go through several rounds of cell division before finally differentiating into a mature cell. It is believed that the molecular distinction between symmetric and asymmetric divisions lies in differential segregation of cell membrane proteins (such as receptors) between the daughter cells.

## SignallingPathways

Adult stem cell research has been focused on uncovering the general molecular mechanisms that control their self-renewal and differentiation.

- *Notch* The Notch pathway has been known to developmental biologists for decades. Its role in control of stem cell proliferation has now been demonstrated for several cell types including haematopoietic, neural and mammary stem cells.
- *Wnt* These developmental pathways are also strongly implicated as stem cell regulators.

## Plasticity/Adult Stem Cell Pluripotency

Discoveries in recent years have suggested that adult stem cells might have the ability to differentiate into cell types from different germ layers. For instance, neural stem cells from the brain, which are derived from ectoderm, can differentiate into ectoderm, mesoderm, and endoderm. Stem cells from the bone marrow, which is derived from mesoderm, can differentiate into liver, lung, GI tract and skin, which are derived from endoderm and mesoderm. This phenomenon is referred to as stem cell transdifferentiation or plasticity. It can be induced by modifying the growth medium when stem cells are cultured *in vitro* or transplanting them to an organ of the body different from the one they were originally isolated from. There is yet no consensus on the prevalence and physiological and therapeutic relevance of stem cell plasticity. More recent findings suggest that pluripotent stem cells may reside in adult tissues in a dormant state. These cells are referred to as "very small embryonic like"—"VSEL"—stem cells, and display pluripotency *in vitro*. As VSEL cells are present in virtually all adult tissues, including lung, brain, kidneys, muscles, and pancreas, co-purification of VSEL cells with other populations of adult stem cells may explain the apparent pluripotency of adult stem cell populations.

## TYPES OF ADULT STEM CELLS

1. **Hematopoietic stem cells** Hematopoietic stem cells are found in the bone marrow and give rise to all the blood cell types.

2. **Mammary stem cells** Mammary stem cells provide the source of cells for growth of the mammary gland during puberty and gestation and play an important role in carcinogenesis of the breast. Mammary stem cells have been isolated from human and mouse tissue as well as from cell lines derived from the mammary gland. Single such cells can give rise to both the luminal and myoepithelial cell types of the gland, and have been shown to have the ability to regenerate the entire organ in mice.

3. **Mesenchymal stem cells** Mesenchymal stem cells (MSCs) are of stromal origin and may differentiate into a variety of tissues. MSCs have been isolated from placenta, adipose tissue, lung, bone marrow and blood, Wharton's jelly from the umbilical cord, and teeth (perivascular niche of dental pulp and periodontal ligament). MSCs are attractive for clinical therapy due to their ability to differentiate, provide trophic support, and modulate innate immune response.

4. **Endothelial stem cells** Endothelial stem cells are one of the three types of multipotent stem cells found in the bone marrow. They are a rare and controversial group with the ability to differentiate into endothelial cells, the cells that line blood vessels.

5. **Neural stem cells** The existence of stem cells in the adult brain has been postulated following the discovery that the process of neurogenesis, the birth of new neurons, continues into adulthood in rats. The presence of stem cells in the mature primate brain was first reported in 1967. It has since been shown that new neurons are generated in adult mice, song birds and primates, including humans. Normally, adult neurogenesis is restricted to two areas of the brain—the subventricular zone, which lines the lateral ventricles, and the dentate gyrus of the hippocampal formation. Although the generation of new neurons in the hippocampus is well established, the presence of true self-renewing stem cells is still under debate. Under certain circumstances, such as following tissue damage in ischemia, neurogenesis can be induced in other brain regions, including the neocortex. Neural stem cells are commonly cultured *in vitro* as so called neurospheres-floating heterogeneous aggregates of cells, containing a large proportion of stem cells. They can be propagated for extended periods of time and differentiated into both neuronal and glia cells, and therefore behave as stem cells. However, some recent studies suggest that this behaviour is induced by the culture conditions in progenitor cells, the progeny of stem cell division that normally undergo a strictly limited number of replication cycles *in vivo*. Furthermore, neurosphere-derived cells do not behave as stem cells when transplanted back into the brain. Neural stem cells may share properties with haematopoietic stem cells (HSCs). Remarkably, when injected into the blood, neurosphere-derived cells differentiate into various cell types of the immune system.

6. **Olfactory adult stem cells** Olfactory adult stem cells have been successfully harvested from the human olfactory mucosa cells, which are found in the lining of the nose and are involved in the sense of smell. If they are given the right chemical environment these cells have the same ability as embryonic stem cells to develop into many different cell types. Olfactory stem cells hold the potential for therapeutic applications and, in contrast to neural stem cells, can be harvested with ease without harm to the patient. This means they can be easily obtained from all individuals, including older patients who might be most in need of stem cell therapies.

7. **Neural crest stem cells** Hair follicles contain two types of stem cells, one of which appears to represent a remnant of the stem cells of the embryonic neural crest. Similar cells have been found in the gastrointestinal tract, sciatic nerve, cardiac outflow tract and spinal and sympathetic ganglia. These cells can generate neurons, Schwann cells, myofibroblast, chondrocytes and melanocytes.

8. **Testicular cells** Multipotent stem cells which are equivalent to embryonic stem cells have been derived form spermatogonial progenitor cells found in the testicles of laboratory mice. This was later confirmed to be of the same capability using cells from the testicles of humans. The extracted stem cells are known as human adult germline stem cells (GSCs). Multipotent stem cells have also been derived from germ cells found in human testicles.

## SOURCES OF STEM CELLS

Pluripotent stem cells, i.e., cells that can give rise to any foetal or adult cell type, can be found in a number of tissues, including umbilical cord blood. Using genetic reprogramming, pluripotent stem cells equivalent to embryonic stem cells have been derived from human adult skin tissue. Other adult stem cells are multipotent, meaning they are restricted in the types of cell they can become, and are generally referred to by their tissue origin (such as mesenchymal stem cell, adipose-derived stem cell, endothelial stem cell, etc.). A great deal of adult stem cell research has focused on investigating their capacity to divide or self-renew indefinitely, and their potential for differentiation. In mice, pluripotent stem cells can be directly generated from adult fibroblast cultures.

### Clinical Applications

Adult stem cell treatments have been used for many years to successfully treat leukaemia and related bone/blood cancers utilizing bone marrow transplants. Early regenerative applications of adult stem cells have focused on intravenous delivery of blood progenitors known as haematopoetic stem cells (HSC's). Other early commercial applications have focused on mesenchymal stem cells (MSC's). For both cell lines, direct injection or placement of cells into a site in need of repair may be the preferred method of treatment, as vascular delivery suffers from a "pulmonary first pass effect" where intravenous injected cells are sequestered in the lungs. In 2008, the first full transplant of a human organ grown from adult stem cells was carried out by Paolo Macchiarini, at the Hospital Clinic of Barcelona on Claudia Castillo, a Colombian female adult whose trachea had collapsed due to tuberculosis. Researchers from the University of Padua, the University of Bristol, and Politecnico di Milano harvested a section of trachea from a donor and stripped off the cells that could cause an immune reaction, leaving a grey trunk of cartilage. This section of trachea was then "seeded" with stem cells taken from Ms.Castillo's bone marrow and a new section of trachea was grown in the laboratory over four days. The new section of trachea was then transplanted into the left main bronchus of the patient. Because the stem cells were harvested from the patient's own bone marrow there was no necessity for anti-rejection (immunosuppressive) medication and when the procedure was reported four months later, the patient's immune system was showing no signs of rejecting the transplant.

## Adult Stem Cell Therapies

The therapeutic potential of adult stem cells is the focus of much scientific research, due to their ability to be harvested from the patient. Similar to embryonic stem cells, adult stem cells have the ability to differentiate into more than one cell type, but unlike the former they are often restricted to certain types or "lineages". The ability of a differentiated stem cell of one lineage to produce cells of another different lineage is called trans-differentiation. Adult stem cell therapies require a stem cell source of the specific lineage needed, and harvesting and/or culturing them up to the numbers required is a challenge.

## Adult Stem Cells and Cancer

In recent years, acceptance of the concept of adult stem cells has increased. There is now a hypothesis that stem cells reside in many adult tissues and that these unique reservoirs of cells not only are responsible for the normal reparative and regenerative processes but are also considered to be a prime target for genetic and epigenetic changes, culminating in many abnormal conditions including cancer.

## MESENCHYMAL STEM CELL

Mesenchymal stem cells, or MSCs, are multipotent stem cells that can differentiate into a variety of cell types, including osteoblasts (bone cells), chondrocytes (cartilage cells) and adipocytes (fat cells). This has been shown in *ex vivo* cultures and *in vitro* or *in vivo*.

### Definition

While the terms mesenchymal stem cell and marrow stromal cell have been used interchangeably, neither term is sufficiently descriptive as discussed below. Mesenchyme is an embryonic connective tissue that is derived from the mesoderm and that differentiates into hematopoietic and connective tissues, whereas MSCs do not differentiate into hematopoietic cells. Stromal cells are connective tissue cells that form the supportive structure in which the functional cells of the tissue reside. While this is an accurate description for one function of the MSCs, the term fails to convey the relatively recently-discovered roles of MSCs in the repair of tissue. Because the cells, called MSCs by many labs today, can encompass multipotent cells derived from other non-marrow tissues, such as umbilical cord blood, adipose tissue, adult muscle or the dental pulp of deciduous baby teeth, yet do not have the capacity to reconstitute an entire organ, the term Multipotent Stromal Cell has been proposed as a better replacement. The youngest, most primitive MSC's can be obtained from the umbilical cord tissue, namely Wharton's jelly and the umbilical cord blood. However the MSC's are found in much higher concentration in the Wharton's jelly compared to the umbilical cord blood, which is a rich source of hematopoietic stem cells. The umbilical cord which is easily obtained after the birth of the new born, is normally thrown away and poses no risk for collection. The umbilical cord MSCs have more primitive properties than other adult MSCs. An extremely rich source for mesenchymal stem cells is the developing tooth bud of the mandibular third molar. While considered multipotent, they may prove to be pluripotent. The stem cells eventually form enamel, dentin, blood vessels, dental pulp, nervous tissues, including a minimum of 29 different unique end organs.

Because of extreme ease in collection at 8–10 years of age before calcification and minimal to no morbidity they will probably constitute a major source for personal banking, research and multiple therapies. These stem cells have been shown capable of producing hepatocytes. Additionally, amniotic fluid has been shown to be a very rich source of stem cells. As many as 1 in 100 cells collected from genetic amniocentesis has been shown to be a pluripotent mesenchymal stem cell. Adipose tissue is one of the richest sources of MSC's when compared to bone marrow. There is 500 times more stem cells in 1 gram of fat than in 1 gram of aspirated bone marrow. Adipose stem cells are currently actively being researched in clinical trials for treatment in a variety of diseases.

## History

In 1924, Russian-born morphologist Alexander A. Maximow used extensive histological findings to identify a singular type of precursor cell within mesenchyme that develops into different types of blood cells. Scientists Ernest A. McCulloch and James E. Till first revealed the clonal nature of marrow cells in the 1960s. An *ex vivo* assay for examining the clonogenic potential of multipotent marrow cells was later reported in the 1970s by Friedenstein and colleagues. In this assay system, stromal cells were referred to as colony-forming unit-fibroblasts (CFU-F). Subsequent experimentation revealed the plasticity of marrow cells. Culturing marrow stromal cells in the presence of osteogenic stimuli such as ascorbic acid, inorganic phosphate, and dexamethasone could promote their differentiation into osteoblasts. In contrast, the addition of transforming growth factor-beta (TGF-b) could induce chondrogenic markers.

## Characteristics

***Morphology*** Mesenchymal stem cells are characterized morphologically by a small cell body with a few cell processes that are long and thin. The cell body contains a large, round nucleus with a prominent nucleolus, which is surrounded by finely dispersed chromatin particles, giving the nucleus a clear appearance. The remainder of the cell body contains a small amount of Golgi apparatus, rough endoplasmic reticulum, mitochondria, and polyribosomes. The cells, which are long and thin, are widely dispersed and the adjacent extracellular matrix is populated by a few reticular fibrils but is devoid of the other types of collagen fibrils.

***Detection*** There is no test that can be performed on a single cell to determine whether that cell is an MSC. There are surface antigens that can be used to isolate a population of cells that have similar self-renewal and differentiation capacities, yet MSCs, as a population, typically do not express all the proposed markers. MSCs have a large capacity for self-renewal while maintaining their multipotency. The standard test to confirm multipotency is differentiation of the cells into osteoblasts, adipocytes and chondrocytes as well as myocytes and neurons. MSCs have been seen to even differentiate into neuron-like cells.

***Immunomodulatory effects*** Numerous studies have demonstrated that human MSC avoid allorecognition, interfere with dendritic cell and T-cell function, and generate a local immunosuppressive microenvironment by secreting cytokines. It has also been shown that the immunomodulatory function of human MSC is enhanced when the cells are exposed to an inflammatory environment characterized by the presence of elevated local interferon-gamma levels.

***Culturing*** The majority of modern culture techniques take a CFU-F approach, where raw unpurified bone marrow or ficoll-purified bone marrow mononuclear cell are plated directly into cell culture plates or flasks. Mesenchymal stem cells, but not red blood cells or hematopoietic progenitors, are adherent to tissue culture plastic within 24 to 48 hours. Other flow cytometry-based methods allow the sorting of bone marrow cells for specific surface markers, such as STRO-1. STRO-1+ cells are generally more homogenous, and have higher rates of adherence and higher rates of proliferation, but the exact differences between STRO-1+ cells and MSCs are not clear. Methods of immunodepletion using techniques such as MACs(Magnetic-activated cell sorting) have also been used in the negative selection of MSCs.

***Clinical use*** The mesenchymal stem cells can be activated and mobilized but the efficiency is very low. For instance, damage to muscles heals very slowly. However, if there were a method of activating the mesenchymal stem cells, then such wounds would heal much faster. Many of the early clinical successes using intravenous transplantation have come in systemic diseases like graft versus host disease and sepsis. However, it is becoming more accepted that diseases involving peripheral tissues, such as inflammatory bowel disease, may be better treated with methods that increase the local concentration of cells. Direct injection or placement of cells into a site in need of repair may be the preferred method of treatment, as vascular delivery suffers from a "pulmonary first pass effect" where intravenous injected cells are sequestered in the lungs.

## CORD BLOOD

Cord blood is a rich source of lifesaving stem cells for your baby. Cord blood is the blood that remains in baby's umbilical cord and placenta after the birth of the baby. This blood has been found to be an extremely rich source of stem cells, the master cells of the human body, that can be used for medical treatments. These stem cells can now be used as an alternative to embryonic, bone marrow and other stem cell types. There are no ethical controversies surrounding stem cells derived from the umbilical cord as the umbilical cord and placenta are usually discarded after birth. Cord blood is a valuable source of stem cells that could be used to treat many life-threatening disease and medical conditions. The umbilical cord blood stem cells are "haemopoetic stem cells" which have a high potency to treat blood and blood related disorders like leukaemia, thalassemia, etc. The value and benefits of umbilical cord blood stem cells are great as stem cells have saved many lives so far, and medical researchers are exploring new uses for umbilical cord blood stem cells in treating diabetes, heart disease and stroke. Cord blood stem cells are useful to regenerate healthy blood and immune system. Umbilical cord blood stem cells offer several advantages over other sources.

1. Stem cells from cord blood are now used to treat over 75 life-threatening diseases.
2. Stem cells from cord blood have successfully treated children and adults.

3. Transplants that use cord blood from a family member (e.g., a sibling) are twice as successful as transplants that use cord blood from a non-relative (e.g., from a public bank).

4. When a child receives his or her own cord blood, there is no risk of rejection.

# 4

# Stem Cell Treatments

Stem cell treatments are a type of intervention strategy that introduces new cells into damaged tissue in order to treat disease or injury. The ability of stem cells to self-renew and give rise to subsequent generations with variable degrees of differentiation capacities, offers significant potential for generation of tissues that can potentially replace diseased and damaged areas in the body, with minimal risk of rejection and side effects. A number of stem cell therapies exist, but most are at experimental stages or costly, with the notable exception of bone marrow transplantation. Medical researchers anticipate that adult and embryonic stem cells will soon be able to treat cancer, Type 1 diabetes mellitus, Parkinson's disease, Huntington's disease, Celiac Disease, cardiac failure, muscle damage and neurological disorders, and many others. Stem cell therapeutics can be applied in the clinical setting, but more research is necessary to understand stem cell behaviour upon transplantation as well as the mechanisms of stem cell interaction with the diseased/injured microenvironment.

## CURRENT TREATMENTS

### Hematopoietic Stem Cell Transplantation

For over 30 years, bone marrow, and more recently, umbilical cord blood stem cells, have been used to treat cancer patients with conditions such as leukaemia and lymphoma. During chemotherapy, most growing cells are killed by the cytotoxic agents. These agents, however, cannot discriminate between the leukaemia or neoplastic cells, and the hematopoietic stem cells within the bone marrow. It is this side effect of conventional chemotherapy strategies that the stem cell transplant attempts to reverse; a donor's healthy bone marrow reintroduces functional stem cells to replace the cells lost in the host's body during treatment.

## Brain Damage

Stroke and traumatic brain injury lead to cell death, characterized by a loss of neurons and oligodendrocytes within the brain. Healthy adult brains contain neural stem cells which divide to maintain general stem cell numbers, or become progenitor cells. In healthy adult animals, progenitor cells migrate within the brain and function primarily to maintain neuron populations for olfaction (the sense of smell). Interestingly, in pregnancy and after injury, this system appears to be regulated by growth factors and can increase the rate at which new brain matter is formed. Although the reparative process appears to initiate following trauma to the brain, substantial recovery is rarely observed in adults. Stem cells may also be used to treat brain degeneration, such as in Parkinson's and Alzheimer's disease.

## Cancer

The development of gene therapy strategies for the treatment of intra-cranial tumours offers much promise, and has shown to be successful in the treatment of some dogs, although research in this area is still at an early stage. Using conventional techniques, brain cancer is difficult to treat because it spreads so rapidly. Researchers at the Harvard Medical School transplanted human neural stem cells into the brain of rodents that received intra-cranial tumours. Within days, the cells migrated into the cancerous area and produced cytosine deaminase, an enzyme that converts a non-toxic pro-drug into a chemotherapeutic agent. As a result, the injected substance was able to reduce the tumour mass by 81 percent. The stem cells neither differentiated nor turned tumorigenic. Some researchers believe that the key to finding a cure for cancer is to inhibit proliferation of cancer stem cells. Accordingly, current cancer treatments are designed to kill cancer cells. However, conventional chemotherapy treatments cannot discriminate between cancerous cells and others. Stem cell therapies may serve as potential treatments for cancer. Essentially, chemotherapy is used to completely destroy the patient's own lymphocytes, and injected stem cells eventually replace the immune system of the patient with that of the healthy donor.

## Spinal Cord Injury

A team of Korean researchers reported on November 25, 2003, that they had transplanted multipotent adult stem cells from umbilical cord blood to a patient suffering from a spinal cord injury and that following the procedure, she could walk on her own, without difficulty. The patient had not been able to stand up for roughly 19 years. For the unprecedented clinical test, the scientists isolated adult stem cells from umbilical cord blood and then injected them into the damaged part of the spinal cord. According to the October 7, 2005 issue of The Week, University of California, Irvine researchers transplanted multipotent human foetal-derived neural stem cells into paralysed mice, resulting in locomotor improvements four months later. The observed recovery was associated with differentiation of transplanted cells into new neurons and oligodendrocytes, the latter of which forms the myelin sheath around axons of the central nervous system, thus insulating neural impulses and facilitating communication with the brain. Researchers at the University of Wisconsin-Madison differentiated human blastocyst stem cells into neural stem cells, then into pre-mature motor neurons, and finally into spinal motor neurons, the cell type that, in the human body, transmits messages from the brain to the spinal cord and subsequently mediates motor function in the periphery.

## Heart Damage

Several clinical trials targeting heart disease have shown that adult stem cell therapy is safe, effective, and equally efficient in treating old and recent infarcts. Adult stem cell therapy for treating heart disease is commercially available. Possible mechanisms of recovery include:

- Generation of heart muscle cells
- Stimulation of growth of new blood vessels to repopulate damaged heart tissue
- Secretion of growth factors

## Haematopoiesis (Blood Cell Formation)

The specificity of the human immune cell allows the human body to defend itself from rapidly adapting antigens. However, the immune system is vulnerable to degradation upon the pathogenesis of disease, and because of the critical role that it plays in overall defence, its degradation is often fatal to the organism as a whole. Diseases of hematopoietic cells are called haematopathology. Research using both hematopoietic adult stem cells and embryonic stem cells has provided insight into the possible mechanisms and methods of treatment for many ailments. Fully mature human red blood cells may be generated *ex vivo* by hematopoietic stem cells (HSCs), which are precursors of red blood cells. In this process, HSCs are grown together with stromal cells, creating an environment that mimics the conditions of bone marrow, the natural site of red blood cell growth. Erythropoietin, a growth factor, is added, coaxing the stem cells to complete terminal differentiation into red blood cells. Further research into this technique should have potential benefits to gene therapy, blood transfusion, and topical medicine.

## Baldness

Hair follicles also contain stem cells which may lead to success in treating baldness through an activation of the stem cells progenitor cells. This treatment is expected to work by activating already existing stem cells on the scalp. Later treatments may be able to simply signal follicle stem cells to give off chemical signals to nearby follicle cells which have shrunk during the aging process. The response to these signals are regeneration and making of healthy hair. Most recently, Dr. Aeron Potter of the University of California has claimed that stem cell therapy led to a significant and visible improvement in follicular hair growth.

## Missing Teeth

In 2004, scientists at King's College London discovered a way to cultivate a complete tooth in mice and were able to grow them in the laboratory. Researchers are confident that this technology can be used to grow live teeth in human patients. In theory, stem cells taken from the patient could be coaxed in the laboratory to turn the tooth bud when they are implanted into the gums to a new tooth. This new tooth is expected to grow within two months. It will fuse with the jawbone and release chemicals that encourage nerves and blood vessels to connect with it. The process is similar to what happens when humans grow their original adult teeth.

## Deafness

Heller has reported success in re-growing cochlea hair cells with the use of embryonic stem cells.

## Blindness and Vision Impairment

Since 2003, researchers have successfully transplanted corneal stem cells into damaged eyes to restore vision. Using embryonic stem cells, scientists are able to grow a thin sheet of totipotent stem cells in the laboratory. When these sheets are transplanted over the damaged cornea, the stem cells stimulate renewed repair, eventually restoring vision. Dr.Sheraz Daya, was able to successfully use adult stem cells obtained from the patient, a relative, or even a cadaver. In April 2005, doctors in the UK transplanted corneal stem cells from an organ donor to the cornea of Deborah Catlyn, a woman who was blinded in one eye when acid was thrown in her eye at a nightclub. The cornea, which is the transparent window of the eye, is a particularly suitable site for transplants. In fact, the first successful human transplant was a corneal transplant. The absence of blood vessels within the cornea makes this area a relatively easy target for transplantation. The majority of corneal transplants carried out today are due to a degenerative disease called keratoconus. In 2009, researchers at the University of Pittsburgh Medical Center demonstrated that stem cells collected from human corneas can restore transparency without provoking a rejection response in mice with corneal damage.

## Amyotrophic Lateral Sclerosis

Stem cells have resulted in significant locomotor improvements in rats with an amyotrophic lateral sclerosis-like disease. In a rodent model that closely mimics the human form of ALS, animals were injected with a virus to kill the spinal cord motor nerves which mediate movement. Animals subsequently received stem cells in the spinal cord. Transplanted cells migrated to the sites of injury, contributed to regeneration of the ablated nerve cells, and restored locomotor function.

## Graft Vs. Host Disease and Crohn's Disease

Clinical trials and therapeutics conducted by Osiris were using their in-development product Prochymal, derived from adult bone marrow. The target disorders of this therapeutic are graft-versus-host disease and Crohn's disease.

## Neural and Behavioural Birth Defects

A team of researchers led by Prof. Joseph Yanai were able to reverse learning deficits in the offspring of pregnant mice who were exposed to heroin and the pesticide organophosphate. This was done by direct neural stem cell transplantation into the brains of the offspring. The recovery was almost 100 percent, as shown both in behavioural tests and objective brain chemistry tests. Behavioural tests and learning scores of the treated mice showed rapid improvement after treatment. On the molecular level, brain chemistry of the treated mice showed rapid improvement after treatment. On the molecular level, brain chemistry of the treated animals was also restored to normal. The researchers discovered that the stem cells worked even in cases where most of the cells died out in the host brain. The scientists found that before they die the neural stem cells succeed in inducing the host brain to produce large numbers of stem cells which repair the damage. These findings, which answered a major question in the stem cell research community, were published in January 2008 in the leading journal, Molecular Psychiatry. Scientists are now developing procedures to administer the neural stem cells in the least invasive way possible or probably via blood vessels, making the therapy practical and clinically feasible. Researchers also plan to work on developing methods to take cells from the patient's own body, turn them into stem cells, and then transplant them back into the patient's blood via the blood stream. Besides the decreased chances of immunological rejection, the approach will also eliminate the controversial ethical issues involved in the use of stem cells from human embryos.

## Diabetes

Diabetes patients lose the function of insulin-producing beta cells within the pancreas. Human embryonic stem cells may be grown in cell culture and stimulated to form insulin-producing cells that can be transplanted into the patient. However, clinical success is highly dependent on the development of the following procedures.

- Transplanted cells should proliferate
- Transplanted cells should differentiate in a site-specific manner
- Transplanted cells should survive in the recipient (prevention of transplant rejection)
- Transplanted cells should integrate within the targeted tissue
- Transplanted cells should integrate into the host circuitry and restore function

## Orthopaedics

Clinical case reports in the treatment of orthopaedic conditions have been reported. To date, the focus in the literature for musculoskeletal care appears to be on mesenchymal stem cells. Centeno *et al.* have published MRI evidence of increased cartilage and meniscus volume in individual human subjects. The results of trials that include a large number of subjects, are yet to be published and a published safety study conducted in a group of 227 patients over a 3–4 year period shows adequate safety and minimal complications associated with mesenchymal cell transplantation. Wakitani has also published a small case series of nine defects in five knees involving surgical transplantation of mesenchymal stem cells with coverage of the treated chondral defects.

## Wound Healing

Stem cells can be used to stimulate the growth of human tissues. In an adult, wounded tissue is most often replaced by scar tissue, which is characterized in the skin by disorganized collagen structure, loss of hair follicles and irregular vascular structure. In the case of wounded foetal tissue, however, wounded tissue is replaced with normal tissue through the activity of stem cells. A possible method for tissue regeneration in adults is to place adult stem cell inside a tissue bed in a wound bed and allow the stem cells to stimulate differentiation in the tissue bed cells. This method elicits a regenerative response more similar to foetal wound healing than adult scar tissue formation.

## Infertility

Culture of human embryonic stem cells in mitotically inactivated porcine ovarian fibroblasts (POF) causes differentiation into germ cells (precursor cells of oocytes and spermatozoa), as evidenced by gene expression analysis. Human embryonic stem cells have been stimulated to form spermatozoon-like cells, yet still slightly damaged or malformed. It could potentially treat azoospermia.

# CLINICAL TRIALS

On January 23, 2009, the US Food and Drug Administration gave clearance to Geron Corporation for the initiation of the first clinical trial of an embryonic stem cell-based therapy on humans. The trial will evaluate the drug GRNOPC1, embryonic stem cell-derived oligodendrocyte progenitor cells, on patients with acute spinal cord injury. As on mid-2010, hundreds of phase III clinical trials involving stem cells have been registered.

# STEM CELL USE IN ANIMALS VETERINARY APPLICATIONS

## Potential Contributions to Veterinary Medicine

Research currently conducted on horses, dogs, and cats can benefit the development of stem cell treatments in veterinary medicine and can target a wide range of injuries and diseases such as myocardial infarction, stroke, tendon and ligament damage, osteoarthritis, osteochondrosis and muscular dystrophy both in large animals, as well as humans. While investigation of cell-based therapeutics generally reflects human medical needs, the high degree of frequency and severity of certain injuries in racehorses has put veterinary medicine at the forefront of this novel regenerative approach. Companion animals can serve as clinically relevant models that closely mimic human disease.

## Development of Regenerative Treatment Models

Veterinary applications of stem cell therapy as a means of tissue regeneration have been largely shaped by research that began with the use of adult-derived mesenchymal stem cells to treat animals with injuries or defects affecting bone, cartilage, ligaments and/or tendons. Mesenchymal stem cells can differentiate into the cells that make up bone, cartilage, tendons, and ligaments (as well as muscle, fat, and possibly other tissues), they have been the main type of stem cells studied in the treatment of diseases affecting these tissues. Mesenchymal stem cells are primarily derived from adipose tissue or bone marrow. Surgical repair of bone fractures in dogs and sheep has demonstrated that engraftment of mesenchymal stem cells derived from a genetically different donor within the same species, termed allogeneic transplantation, does not elicit an immunological response in the recipient animal and can mediate regeneration of bone tissue in major bony fractures and defects. Stem cells can speed up bone repair in fractures/defects that would normally require extensive grafting, suggesting that mesenchymal stem cell use may provide a useful alternative to conventional grafting techniques. Treating tendon and ligament injuries in horses using stem cells, whether derived from adipose tissue or bone-marrow, has support in the veterinary literature. While further studies are necessary to fully characterize the use of cell-based therapeutics for treatment of bone fractures, stem cells are thought to mediate repair via five primary mechanisms: 1) providing an anti-inflammatory effect, 2) homing to damaged tissues and recruiting other cells, such as endothelial progenitor cells, that are necessary for tissue growth, 3) supporting tissue remodelling over scar formation, 4) inhibiting apoptosis and 5) differentiating into bone, cartilage, tendon, and ligament tissue.

Currently available treatments for horses and dogs suffering from orthopaedic conditions

Autologous or allogeneic stem cells are currently used as an adjunctive therapy in the surgical repair of some types of fractures in dogs and horses. Autologous stem cell-based treatments for ligament injury, tendon injury, osteoarthritis, osteochondrosis, and sub-chondral bone cysts have been commercially available to practicing veterinarians to treat horses since 2003 in the United States and since 2006 in the United Kingdom. Autologous stem-cell based treatments for tendon injury, ligament injury, and osteoarthritis in dogs have been available to veterinarians in the United States since 2005. Over 3000 privately-owned horses and dogs have been treated with autologous adipose-derived stem cells. The efficacy of these treatments has been shown in double-blind clinical trials for dogs with osteoarthritis of the hip and elbow and horses with tendon damage.

### Developments in Stem Cell Treatments in Veterinary Internal Medicine

Currently, research is being conducted to develop stem cell treatments for: 1) horses suffering from COPD, neurologic disease, and laminitis; and 2) dogs and cats suffering from heart disease, liver disease, kidney disease, neurologic disease, and immune-mediated disorders.

## SIGNIFICANCE OF STEM CELL MICROENVIRONMENTS

The microenvironment into which stem cells are transplanted significantly alters the capacity of engrafted cells for recovery and repair. The microenvironment provides growth factors and other chemical signals that guide appropriate differentiation of transplanted cell populations and direct transplanted cells to sites of trauma or disease. Repair and recovery can then be mediated via three primary mechanisms: 1) formation and/or recruitment of new blood cells to the damaged region; 2) prevention of programmed cell death or apoptosis; and 3) suppression of inflammation. To further enrich blood supply to the damaged areas, and consequently promote tissue regeneration, platelet rich plasma could be used in conjunction with stem cell transplantation.

## SOURCES OF AUTOLOGOUS (PATIENT-DERIVED) STEM CELLS

Autologous stem cells intended for regenerative therapy are generally isolated either from the patient's bone marrow or from adipose tissue. The number of stem cells transplanted into damaged tissue may alter efficacy of treatment. Stem cells derived from bone marrow aspirates, for instance, are cultured in specialized laboratories for expansion to millions of cells. Although adipose-derived tissue also requires processing prior to use, the culturing methodology for adipose-derived stem cells is not as extensive as that for bone marrow-derived cells. While it is thought that bone-marrow derived stem cells are preferred for bone, cartilage, ligament, and tendon repair others believe that the less challenging collection techniques and the multi-cellular microenvironment already present in adipose-derived stem cell fractions make the latter the preferred source for autologous transplantation.

# 5

# Stem Cells and Nervous System

## STEM CELL TREATMENT FOR MS

Multiple sclerosis is an autoimmune disorder mainly affecting the young adults and is characterized by destruction of myelin in the central nervous system. The autoimmune attack includes auto-reactive lymphocytes and local inflammatory responses that cause demyelination and oligodendrocyte death followed by accumulating axonal damage and axonal loss. Pathologic findings include multiple sharply demarcated areas of demyelination throughout the white and grey matter of the central nervous system.

## CLINICAL MANIFESTATIONS OF MS

The clinical manifestations include local anaesthesia and anergia of one or more parts of the extremities; pricking pain; cold; diminution of vision; blurry-eyed; double vision and so on. Some patients have the appearance of acute or progressive slight spastic paraplegia and anaesthesia. MS have the potential to recur. On recurrence the symptoms are same as before or new symptoms could develop. The usual symptoms may include:

1. Paralysis, acratia or a feeling of ponderosity
2. Sensory disturbance

3. Dysequilibrium
4. Vision defects, double vision
5. Progressive atony
6. Acute myelitis
7. Pain
8. Sense of girdle
9. Cold in localized regions of the body
10. Dysarthria
11. Lhermitte's sign
12. Paroxysmal pruritus.

## STEM CELL TREATMENT

The treatment has been programmed to combine both nerve stem cell activation and stimulation treatment and stem cell injections. The treatment includes immunotherapy to improve and adjust the function of the patient's immune system, and stem cell implantation, a "cocktail" of medications and neural nutrition materials given through IV, biofeedback treatment and comprehensive rehabilitation. These methods can maximally regulate the states of the immune system, improve symptoms such as eyesight problems, muscular weakness, sensory disturbance and paralysis of the body. The curative effect depends on the patient's medical condition. In general, younger MS patients have shown greater benefits from the stem cells treatment as the disease was not far advanced. Some patients had their MRI scans change to normal after half a year and this does not mean that the disease is cured completely. Current treatment has only proven that it can slow the progression of this active disease.

### Autogenous Stem Cells Activation and Proliferation

Under non-invasive procedures, through daily IV of an individually customized cocktail of neurotrophic medicines we activate the patient´s own stem cells and thus turn these cells into nerve precursor cells. These have signalization and migrate to lesions along the nerve developing-chord.

## Stem Cell Injections

Before the injection of stem cells, immunosuppressors are used to stop the cause of MS. As an autoimmune disease, the main treatment aims to restrain the process of inflammatory demyelination. Hence the immune treatment consists of corticosteroid medicines such as methylpredisolone, dexamethasone and prednisone, etc., to adjust the function of the patient's immune system. Interferon is used to inhibit the function of their immune cells. The dosage of these medications is very important thereby the stem cells treatment will not interfere with the interferon. With the cooperation of the process of immunoloregulation, interferon treatment and stem cells treatment, the patients can get a good prognosis.

***Collection and injection of autogenous stem cells*** The potential stem cells are collected from bone marrow and cultured so as to proliferate these cells into mesenchymal stem cells (MSC). This procedure needs 3–4 weeks to get the required amount of MSC (4,000,000). The mesenchymal stem cells (MSC) are implanted through the SCF (spinal cord fluid) by means of a lumbar puncture. Each injection has 5,000,000 stem cells. As this procedure is short and simple, it is usually done with local anaesthesia. Moreover to prevent any risk of infection, the procedure is done in a strictly sterile environment—the operating room—by an experienced neurology doctor only. The advantage of this treatment is that there are no immune side effects since the stem cells are from the patient's own body; therefore, there is no need to use an immunosuppressant to restrain the immune rejection of these stem cells. The self stem cell activation program, as well as Chinese traditional medicine and a comprehensive rehabilitation program are given to the patients at the time of cell culture.

***Neural stem cells injection*** This treatment is given to patients who suffer from some inherited disease that is a genetic defect and in elderly patients where the autogenous stem cells are unfit for the stem cells treatment. Neural stem cells are multipotent stem cells that can develop into a variety of cell types, including neurons and neural ganglia cells. Neural stem cells can be acquired conveniently. At present the neural stem cells come from Peking University's Stem Cells Research Centre, China, where the cells are rigorously tested with strict regulations. So the advantages are: Firstly, they are convenient and time saving because they do not have to be cultured for a long time. Secondly, the neural stem cells are safe and secure in clinical therapy. In addition, these stem cells have good biological activity and the immune response is minimal (immunological rejection) within the patients. The method of stem cells injection is the same as autogenous stem cells injection.

***The side effects of stem cells treatment*** The procedure can cause minor side effects for 2 to 3 days after the stem cells injection, such as fever, sleepiness, headaches, dizziness and fatigue. Some patients do not suffer any side effects.

## Rehabilitation

Rehabilitation therapy is very much a necessary as part of the regimen. The stem cell mobilization technique is combined with signal induction in the rehabilitation training procedure which can significantly improve the patient's neural function. The Rehabilitation Centres use different kinds of rehabilitation methods such as Physical Therapy, Occupational Therapy, Speech Therapy and Traditional Chinese Medicine (TCM).

## Treatment Period

The recommended period of treatment is about 5–6 weeks.

# KINDS OF STEM CELL

## Autogenous Stem Cell/Neural Stem Cell

Before treatment with the stem cells, the patient has to be evaluated, especially the clinical condition of their nervous system and immune system, so as to plan the stem cell treatment according to the patient's condition. In general autogenous stem cells treatment is advised because there are no immunological side effects since those stem cells are from the patient's own body; but if the patient is too old, neural stem cells are introduced as these stem cells are biologically active and possess the ability to differentiate easily.

***Contradiction*** Pituitary tumour and adrenocortical hyperplasia occurs and hence the patient is under observation.

### *Medical plan procedure*

1. The immunologic function of the patient is improved and adjusted to end the aetiology of this disease.
2. Medication to improve the internal microenvironment so that it is suitable for the subsistence of stem cells.
3. Stem cells implantation by subarachnoid cavity injection.
4. Signal inductive treatment to contribute to stem cells fix planting in a particular location. This is according to the theory of immunology and endocrinology.
5. Prompt stem cells differentiation to help the neural precursor cells differentiate and develop into neurons, oligodendrocytes and neurogliocytes.

## STEM CELL TREATMENT FOR CEREBRAL PALSY

A comprehensive program of therapy has been developed for Cerebral Palsy patients by combining both nerve stem cell activation and stimulation treatment, and stem cell injections. This combined treatment has so far proven itself to be the most effective means in achieving a high level of recovery. The treatment includes stem cell implantation, a "cocktail" of medications and brain nutrition materials infused intravenously, biofeedback treatment and comprehensive rehabilitation. All of the above methods can maximally improve functions such as sensibility, mobility, language and intelligence. The key to the success of this treatment method is the overall customized treatment regimen for each patient.

***Self stem cells activation and proliferation*** Using a non-invasive procedure, through daily IV of an individually customized cocktail of neurotrophic medications, the patient's own stem cells are activated and as a result, these cells are customized into nerve precursor cells. These cells have signalization and migrate to lesions along the nerve developing-chord.

***Neural stem cells injections*** The neural stem cells are extracted from a foetus;

A safe stem cell implantation procedure for Cerebral palsy patients is as follows;

The procedure includes sub-arachnoid surgery, spinal cord fluid stem cell injections, and an injection through a lumbar puncture into the spinal cord fluid. This technique is used in the following cases where, the patient cannot cooperate properly for going through stereotactic brain surgery. The target in the brain is not clear enough to use the brain injection. This procedure takes between 5 and 10 minutes. It is a simple and safe procedure that has almost no side effects or risks. This is performed in the operation theatre by stem cell specialists. Typically, 3 or 4 injections of 5 million cells each are given to the patient.

## Rehabilitation

Rehabilitation therapy is a very necessary part of the whole treatment regimen. In the past 20 years, nerve system rehabilitation, in terms of motor control theory, has been greatly developed. The research results indicate that brain injury is an anatomical injury and it does not necessarily mean functional loss. Modern Nerve Development Theory (NDT) focuses on what functions remain and what functions can be regained through treatment and/or training. The adult stem cell mobilization technique is an inclusion of signal induction in the rehabilitation training procedure which can significantly improve brain function. There are different kinds of rehabilitation methods including Physical Therapy, Occupational Therapy and Speech Therapy.

# THE PROCEDURES

## Self Stem Cell Activation and Proliferation Program

This program usually requires a 5-week stay in the hospital. The health and the details of the patient's condition, the neurological function and the pathological change in the brain tissues are established by the medical team. The medical plan is designed for self-stem cell original activation and proliferation treatment. To drive off the trigger factors of the stroke, the blood vessel wall, the blood ingredients and the blood flow are treated to improve the brain microenvironment to help the stem cells proliferate. The self-activation and proliferation treatment is then started. Based on the Neuro-endocrine Immune Network Theory, patient is treated by "waking up" the sleeping stem cells through oral medication. Different medications are used during different processes (proliferation-inducement-differentiation), the medications are always adjusted and combined; therefore, it is termed as "cocktail treatment". Throughout the treatment period we will be following up and evaluating the neurological function and improvement. After the first treatment cycle is complete, 2 to 5 months of rest is recommended before starting the second treatment.

## Stem Cell Implantation through the Cerebral Spinal Fluid (Lumbar Puncture)

Most cerebral palsy patients have global brain damage. That is why the stem cell implantation through the spinal cord fluid is recommended. The stem cells will migrate to the damaged areas in the brain through the CSF. This procedure is short and simple and done with local anaesthesia. In order to prevent even the minor risk of infection and maintain a strictly sterile environment, the procedure is done in the operating room by doctors only.

***Side effects*** The procedure can cause minor side effects for 2 to 3 days after the operation including fever, headaches, dizziness and fatigue. These side-effects are expected. Many patients do not suffer any side effects at all.

***Treatment period*** The recommended time period for the treatment is 5 weeks, but there should be improvement after the first month. The patient can decide whether or not to continue with the second month's treatment at the end of the first month.

## PARKINSON'S DISEASE (TREMOR ARTUUM)

This is the most common disease with the central nervous system denatured; most of the patients are middle-aged or elderly. The substantia nigra cells are located in the mesencephalon area and when the pathology changes, this leads to dopamine synthesis reduction, inhibition of the functioning of acetylocholine, and the relative increase of acetylcholine excitation. These imbalances result in "tremor artuum". The clinical signs and symptoms include increased tension in all four limbs, trembling of the hands, balanced function reduction, gravity anteversion and "flustered gait". Therefore, in recovery training, the primary focus is on alleviating the trembling of the patient's four limbs, increasing the patient's muscular strength in the upper limbs, improving muscular control of the upper limbs, improving balance through the use of the balance plate, both ball and other tools, and getting to the point where the patient's balance is good enough without the need for support. Furthermore, the patient's gait is continuously corrected throughout the course of treatment. The patient is also trained to restore oscillation arm functioning when walking.

## MULTIPLE SYSTEM ATROPHY

This is multiple spot nervous system denatured disease; it affects the extrapyramidal system, pyramidal system, the cerebellum and the parasympathetic system, and so on. Multiple system atrophy involves corpus striatum-substantia nigral denaturation which mainly presents with Parkinson's type symptoms, Chinese olive-pons-cerebellum atrophy as well as parasympathetic system's functional barrier, which mainly presents with Shy-Drager syndrome. MSA patients' clinical symptoms are similar to those of Parkinson's patients. Hence the treatment modality is also quite similar. Some recreational activities according to the patient's particular case are arranged but we recognize that some MSA patients may be presented with body posture hypotension, which is different with Parkinson's patients.

## AMYOTROPHIC LATERAL SCLEROSIS

This is the most common type of motor neuron disease. It affects the ventricolumna cells, brain stem, coordination, movement, nerve nucleus and pyramidal tract. There will initially be damage to the upper motor neurons or lower motor neurons, eventually leading to the development of lower motor neuronal lesions. Patient's clinical presentation includes muscular atrophy, decreased muscular strength; some patients may also have anarthria, difficulty with swallowing and chewing, tongue amyotrophy and affective disturbance. The treatment is focused mainly on enhancing the patient's muscular strength through support training as well as methods designed to restore language and swallowing functioning. All three are characterized as nervous system diseases, and the cause and mechanisms of all three diseases is not explicit. During the recovery treatment, besides the symptomatic treatment, communication with the patient is extremely important. All three of these diseases may have a relation to long-term nutrition imbalance; hence it is important to plan each patient's diet. Our comprehensive treatment plan is focused on improving the overall quality of life for each patient.

## STEM CELL TREATMENT FOR ENCEPHALITIS

Encephalitis is a kind of infection of the central nervous system caused by a virus, germ or parasite. Patients will have sequela of encephalitis because of the damage/degeneration of the nervous system. The main symptoms include motor disturbance, dysphasia, seizure and intelligence deficit. Many patients have been treated during the post infectious stage of encephalitis; most of them had obvious improvements.

For patients who are in the stage of post infectious encephalitis, a unique program combining several therapies is provided:

Before stem cells treatment, a complete examination of the patient's condition is done by the neurologist including organ functioning and evaluation of the nervous system.

Medication is provided to adjust the condition of the patient's immune system. According to the patient's clinical condition, an effective medication is used to adjust the immune function, nourish neurons and clear away the internal microenvironment to protect the neurons and help the growing of stem cells.

***Stem Cells Activation and Proliferation Treatment*** Patients receive a daily IV of neurotrophic factors and medicines used to stimulate the production of the body's own stem cells.

## Rehabilitation

The rehabilitation program is an integral part of stem cell therapy. The treatment includes daily physiotherapy and occupational therapy sessions, and a complementary Traditional Chinese Medicine treatment (professional acupuncture and massage).

## STEM CELL IMPLANTATION

Neural stem cell injections are delivered via lumbar puncture into the cerebral spinal fluid in four separate injections. Those stem cells will grow into neurons and rebuild new body neural innervations, at the same time, the new neurons can provide nourishment to the peripheral nerves and nerve roots, fixing the damage to the neural system.

## Program Length

The program length is about 5–7 weeks. Self Bone Marrow is an alternative treatment program that utilizes stem cells sourced from the patient's own body. Self bone marrow is extracted from the patient's hip, then cultured in the lab and induced into neural stem cells. Finally, the cultured cells are injected back via the spinal cord fluid in 2–4 injections. Self bone marrow stem cell implantation has no problem of rejection.

## Side Effects

Possible side effects include headache, fever, weakness and dizziness. Side effects can persist between the first 24 hours and 3 days after the injection. These side effects are temporary, natural, and have no serious threat to the patient's health. A stronger rejection reaction is rare but possible.

# STEM CELL TREATMENT FOR STROKES

A stroke is a leading cause of severe, long-term disability worldwide. 50–70% of survivors suffer from a range of symptoms including weakness, numbness, paralysis, speech difficulties, confusion and other problems. A stroke is an injury to the brain due to the interruption of the blood supply, which causes destruction of a portion of brain tissue. One of the consequences that occur after a stroke is a severe burning pain on the side of the body affected by the stroke. Post-stroke pain is also referred to as thalamic pain, neurogenic pain or central pain. Neurogenic pain or central pain syndrome occurs frequently following strokes on the right side of the brain, affecting the left side of the body. So far more than 50 stroke patients have gone through "Self Stem Cells Activation and Proliferation Program," using the adult nerve stem cell mobilization technique to bring the remaining nerve cells of the post stroke patient into action. Recently a comprehensive program is developed to treat post stroke patients by combining both neural stem cell activation and stimulation, and stem cells injections. This combined treatment has proven to be the most effective method. The treatment includes stem cell implantation, medication procedure, surgical procedure, biofeedback treatment and comprehensive rehabilitation. This combination therapy improves functions such as sensibility, mobility, language and intelligence.

## SELF STEM CELLS ACTIVATION AND PROLIFERATION

A non-invasive procedure is used and through a daily IV of an individually customized cocktail of neurotrophic medications the patient's own stem cells is activated and turned into nerve precursor cells. These cells have signalization and migrate to lesions along the nerve developing-chord.

## NEURAL STEM CELLS INJECTIONS

Neural stem cells extracted from a foetus have the advantage of being specialized naturally into neural cells and therefore there is no need to induce them into neural cells by artificial means. These stem cells can be injected by a lumbar puncture or by Sub-arachnoid Surgery. This technique is used in the following cases where the patient cannot co-operate properly for going through stereotactic brain surgery. The target in the brain is not clear enough or is too big to use the brain injection. The patient has a fear of brain surgery and prefers to have the lumbar puncture injection. This procedure takes between 5 and 10 minutes. It is a simple and safe procedure and that has almost no side effects or risks. This procedure is done in the operating room and is performed by stem cell specialists. Typically, between 3 and 4 injections of 5,000,000 cells each are given to the patient.

## Rehabilitation

Rehabilitation therapy is a crucial part of the entire treatment regimen. As per the motor control theory, the nerve system rehabilitation indicates that the brain injury is an anatomical injury and it does not necessarily mean functional loss. Modern Nerve Development Theory (NDT) stresses what functions remain and what functions can be regained through treatment. The adult stem cell mobilization technique is combined with signal induction in the rehabilitation training procedure which can significantly improve brain function. The rehabilitation methods include Physical Therapy, Occupational Therapy, Speech Therapy and Traditional Chinese Medicine.

# THE PROCEDURES

## Self stem cell activation and proliferation program

This program usually requires five to six weeks of stay in hospital. A thorough evaluation of the neurological function and understanding of the lesion or pathological change in the brain tissues is established by the medical team. According to the condition of the patient, the treatment is designed for the self stem cell original activation and proliferation treatment. Then, to drive off the trigger factors of the stroke, the blood vessel wall, the blood ingredients and the blood flow are treated, thereby improving the brain microenvironment to help the stem cells proliferate. The self activation and proliferation treatment is done. Based on the Neuro-endocrine Immune Network Theory, this method 'wakes up' the sleeping stem cells through oral medication. We use different medications during different processes (proliferation-inducement-differentiation). The medications are always adjusted and combined. Hence this is termed as "cocktail treatment". Throughout the treatment period the progress of the neurological function and improvements are assessed. After the first treatment cycle is complete, 2 to 5 months of rest are recommended before starting the second treatment.

# 6

# Role of Stem Cells in Spinal Cord Injury

Spinal cord injury is defined as damage to the spinal cord that results in partial or complete loss of function, whether temporary or permanent. The three most common causes of spinal cord injury are direct trauma, compression by disc herniation, and damage caused by occluded spinal arteries. The range of symptoms that result from a spinal cord injury is largely determined by the size and location of the lesion and most patients experience chronic pain accompanied by lifelong heart and lung complications. Current therapies focus on physical rehabilitation and counseling to deal with the emotional frustration of disability, but nothing can be done to regenerate the spinal cord. As such, research for treatments targets four main concepts: limiting the damage, neuroreconstruction, stimulating regrowth of neurons, and retraining neural circuits to restore body functions. Stem cells have shown great promise and potential for restoration of the damaged spine.

## MECHANISM OF INJURY

Most spinal cord injuries arise from a direct traumatic blow to the spine. Motor vehicle accidents, violent encounters, falls, and sports accidents cause the majority of these injuries. The destructive process, however, occurs in two steps. First, the initiating event dislocates or fragments the vertebrae, sending shards of bone into spinal cord tissue. This causes the obvious lesion and then a slew of biochemical and cellular events alter the blood flow. Leakage from the comprised blood vessels exacerbates compression of the spinal cord and inflammatory agents trigger invasion of immune system cells, causing even more cellular and fluid accumulation. The release of inflammatory signaling molecules or cytokines induces morphologic and functional changes in the supporting cells in the region.

## CLASSIFICATION OF SPINAL CORD INJURIES

After the injury has been stabilized, a complete neurological exam is performed to determine the true degree of damage. Injuries may be classified as either paraplegic (affecting only the lower extremities) or tetraplegic (affecting all four limbs, also known as quadriplegic). The American Spinal Injury Association has developed a more specific classification system to help predict the likelihood and extent of recovery. The categories range from an A-level, at which no motor or sensory functions are retained below the level of injury (also referred to as a complete injury) to E-level injuries, in which motor and sensory functions are normal. Categories B through D are incomplete injuries, in which varying degrees of sensation and motor functions are preserved. Although some people are fortunate enough to preserve their motor function, almost all will experience cardiovascular and pulmonary complications—the most common causes of death following a spinal cord injury. These include pneumonia, arrhythmia, blood pressure dysregulation, increased tendency for blood clots, and possible breathing problems (if damage occurs above the phrenic nerve, which innervates the diaphragm). Chronic pain, bladder/bowel problems, and sexual dysfunction are also common among spinal cord injury survivors.

## TREATMENTS

The severity of disability following a spinal cord injury is directly correlated with the size of the lesion and the effects of the secondary inflammatory cascade. Hence the treatment modalities immediately following trauma include immobilization of the spine (during transport), use of steroids (i.e., methylprednisone) to decrease inflammation, and spinal realignment via traction. Following stabilization, not much can be done except physical rehabilitation, counseling, and management of chronic symptoms. Though no effective restorative treatment exists as of right now, stem cells are a promising source for a cure.

## RESEARCH

Current research in the field of spinal cord injuries is directed towards four principles: limiting damage, promoting regrowth of neurons, inducing new neuron growth, and retraining neural circuits. This is divided itself into two main tracts of study. One involves the manipulation of biochemical signals to create a growth-promoting, damage-limiting microenvironment, and the other works to replace damaged cells. For the latter, stem cell transplantations are an obvious choice for neuro-reconstruction, but they also contain secondary potentials for producing growth-supporting factors thereby forming a cellular scaffold to help regenerating axons bridge across the scarred spinal tissue. The advantage of transplanting undifferentiated stem cells allows specific adaptation to the microenvironment and subsequent production of appropriate cell populations. There have already been successful proof-of-principle studies performed using animal models. However, before developing a clinically viable stem cell treatment, several obstacles lie in the way. (1) ease of obtaining a reliable (2) noncontroversial proof of stem cells (3) optimization to improve graft integrations (4) decreasing the severity of side effects. A less significant hindrance is that treatment for most spinal cord injuries must be delayed until the acute inflammation is resolved. An Australian group demonstrated the ability of supporting cells (olfactory ensheathing cells) from a rat nasal cavity to facilitate partial regeneration of a completely severed rat spinal cord. As nasal biopsies are a relatively simple, low-risk procedure, they could provide an easily accessible source of transplantation material. Also, in consideration of the likely delay seen in clinical settings, the researchers incorporated a four-week wait between cord transaction and transplantation. Other stem cell types that have been targeted for neuro-reconstructive potential include human embryonic stem cells, mesenchymal stem cells, and neural stem cells. Embryonic stem cells are a logical choice because they allow the widest range of differentiation with almost infinite reproduction. Mesenchymal stem cells (obtained from bone marrow or umbilical cord blood) have demonstrated some degree of success at differentiation into neuronal components. Neural stem cells, which can be either exogenous (from an outside source) or endogenous (intrinsic to the adult spinal cord), are considered to be easier to differentiate because of their being committed to a neural fate. As with all cases of transplantation, there is an obvious risk for infection or rejection by the recipient's immune system. With implantation into the nervous system, there is the additional risk of developing syringomyelia, a condition in which a cyst grows within the spinal cord, thus damaging more tissue and worsening the situation. Other serious adverse effects include uncontrolled cancerous growth, migration of the implant, or neural dysfunction arising from improper connections between reconstructed nerves. To overcome these disadvantages, the nuances of stem cell types need to be elicited, an accessible source of stem cells that can proliferate enough to create viable transplants needs to be identified, and an understanding of how to successfully manipulate the signals governing cell differentiation needs to be gained. Hence the researchers should strive for the complete restoration of function following a spinal cord injury.

## SPORTS INJURIES

Sports injuries are defined as any type of injury that arises as a result of playing or practicing any kind of sport. As there are a wide range of sports, the range of injuries caused by them is also very wide. Motor car racing, for example, can lead to massive trauma and wounding from high-speed collisions, whereas boxing or other contact sports can lead to brain damage after repeated striking of the head. In other cases, sports injuries are comparatively minor, focusing on muscle strains and tears, bone injuries, and cuts and abrasions. Sports injuries and sports medicine and a wide range of different branches of medicine and a wide range of possible interventions:

Cardiology, orthopaedic surgery, nutrition, and physical therapy are among the fields covered. Sports medicine has pointed out that repeated incidents of concussion among male athletes, for example, have been found to be linked to clinical depression. Meanwhile, drugs that are taken to bulk up muscles may actually lead to more injuries, because of the interaction between muscles and bones and joints. Well-known actor Christopher Reeve suffered a serious spinal injury while taking part in equestrian sports and subsequently became a powerful voice in favour of the kind of stem cell research that might offer some treatment for him and for people in similar situations in the future. Reeve-Irvine Research Center has come to be an important source of new research, and indeed advocacy, in support of the cause.

## FUTURE TREATMENT

Minimally invasive surgery techniques have in part resulted from sports injuries. New techniques have yet to be created or developed for human use. For example, creating healthy regimes with respect to nutrition counseling, physical therapy and exercise physiology. However, the areas in which stem cell research has provided actionable areas of medical treatment lie mostly in the repair of worn out (over-used) muscular body parts or in rapid healing of wounds. Research has demonstrated that wounds may be treated more quickly and more effectively by using stem cells from hair follicles in conjunction with bone marrow-derived stem cells to repair wounded epidermis. Research also suggests depositing stem cells, perhaps from umbilical cord blood, either by the athlete or by offspring into a suitable tissue bank so that patient's own tissue is available which sidesteps the rejection of foreign tissue. Stem cell research in the future may be caused to grow into replacements for knee or elbow cartilage, for example, which are areas of the body that can be of considerable importance (and financial relevance) in an athlete's career. Other forms of body tissue might also be feasibly replaced in the future.

Stanford University is located in California; it was founded in 1885 and opened in 1891, and the Stanford Medical Center opened in 1959. Stanford is known for its fundamental stem cell research as well as clinical trials to translate science into medical treatment for human disease, including heart disease and neurological disease. Stanford researchers have developed technology necessary for stem cell research. In the 1980s, researchers developed a high-speed cell sorter and used the technology to isolate stem cells. In the 1990s, the development of microarray technology made it possible for researchers to assess differences in genetic expression throughout the stages of development and distinguish between normal and cancerous tissues. Researchers in the genetic department developed the fluorescence activated cell sorter for identification and extraction of pure cells from tissue samples. The Stanford Institute for Stem Cell Biology and Regenerative Medicine was formed to focus research on creating new embryonic stem cell lines, organ and tissue regeneration and cancer cure/therapy through targeting the cancer stem cell.

## Embryonic Stem Cell Line Creation and Research at Stanford

Researchers are isolating heart and blood-forming stem cells derived from human embryonic stem cells and have created four new embryonic stem cell lines by In-Vitro Fertilization (IVF) embryos. A California Institute for Regenerative Medicine grant provides researchers with the opportunity to expand using embryos with genetic likelihood of developing Down syndrome, cystic fibrosis, or muscular dystrophy and to isolate heart and blood stem cells from embryonic stem cells. In addition to those lines from IVF embryos, researchers hope to create stem cells genetically identical to a specific person by using somatic cell nuclear transfer (SCNT), a process that removes the egg nucleus and replaces it with a nucleus from an adult cell, then stimulates the egg to develop for the removal of genetically identical embryonic stem cells. This technique could be used to either create cells for therapy or for further research on the diseases of the human donor. Because of the ability to work with the IVF clinic to obtain rejected eggs and increase knowledge, the researchers hope to create human stem cell lines through SCNT. Additional research includes the creation of human embryonic stem cell lines and the derivation of different cell types from both new and existing cell lines, as well as understanding and improving the techniques for transforming adult skin cells into cells that act like embryonic stem cells.

## Organ and Tissue Regeneration

In 1987, Stanford researchers identified blood-forming stem cells and identified brain, muscle, and other tissue stem cells. This research includes usage of adult stem cells in lung, peripheral nervous system, skin, liver, and pancreas, with the possibility of regenerating organs and tissues. Researchers worked with Stanford's physicians and transplant teams to perform clinical transplantation of blood stem cells and brain stem cells to regenerate tissue. California Institute for Regenerative Medicine grant money has been awarded to researchers who plan to generate inner ear sensory cells from human embryonic stem cells to be used in tissue grafting as a hope for treating deafness and to engineer cardiovascular tissue from human embryonic stem cells.

## Cancer Stem Cell Research

Researchers at Stanford were the first to identify and isolate human leukaemia and human breast cancer stem cells. Since only a limited number of cancer stem cells can proliferate without control, researchers at Stanford plan to continue their work on possibilities for treating cancer by stopping these cancer stem cells from growing out of control and destroying tissues and organs. The researchers are currently working on human brain, ovarian, melanoma, and bladder stem cells. They have used stem cell research to understand how cancer cells regenerate. Using cancer stem cells, the researchers identify cancer genes within stem cells and develop clinical approaches to destroy cancer stem cells as a treatment for cancer. Their work in genetics is focused on creating stem cells through nuclear transfer to study the diseases and develop treatment options for inherited diseases, cancer, heart disease, diabetes, autoimmune diseases, and neurodegenerative disorders. Scientists and specialists from a variety of related scientific fields including cell biology, nanotechnology, applied physics, bioengineering, biocomputation, chemical biology, and tissue engineering pave ways to advance regenerative technology by using stem cells to repair diseased or damaged tissues and organs.

## Stem Cells, Inc.

STEMCELLS, INC., is a California-based company dedicated to diseases of the brain, liver, and pancreas. The company works with tissue-derived stem cells with the goal of using these cells for regenerative therapy and holds human neural stem cell, with candidates for both hepatic and pancreatic stem cells. The company defines stem cells as those cells that have the ability to differentiate into any other cell type and that can divide sustainably. Furthermore, those cells that have risen from stem cells but can still differentiate into multiple cell types are called progenitor cells.

Stem cells are harvested from tissues such as the brain, liver or pancreas through a series of purifications. The ultimate goal is to use these tissue derived stem cells for therapeutic purposes, such as to restore the health of a person suffering from diabetes, chronic liver disease, Alzheimer's disease or any other degenerative disorder, as well as spinal cord injury. Stem cells look forward to use chemicals produced and secreted by stem cells for therapeutic purposes, as these chemicals are typically involved in growth and differentiation. The three chief areas of research being performed by stem cells are the brain, the liver, and the pancreas. In the year 1997, scientists developed a method for culturing human stem cells for the central nervous system (CNS). The CNS consists of the brain and spinal cord. These CNS stem cells can be reproducibly grown in culture for several generations and can become neurons or glia—the support cells of the brain. The two chief glial types in the CNS are astrocytes and oligodendrocytes; CNS stem cells can differentiate into neurons or either glial types. These stem cells were transplanted into mouse brains and found to differentiate into various cell types depending on the region into which they were transplanted. Method for the purification of human CNS stem cells from the brain has been developed using monoclonal antibodies. This technique allows a pure sample of stem cells to be obtained, which is critical for more advanced therapeutics. Target diseases include Alzheimer's disease; CNS demyelination disorders such as multiple sclerosis; lysosomal storage disorders such as neuronal ceroid lipofuscinosis (Batten disease), which are particularly disastrous to neurons and spinal cord injury. In addition to disease therapy, these stem cells also could be used for testing drugs for toxic side effects, as well as for genetic analysis to achieve a greater understanding of genes involved in growth, differentiation, and repair. In the hepatic stem cell research area, these are used to treat cancer, genetic liver diseases, and hepatitis. One specific liver disease is tyrosinemia type 1, which is fatal. The mouse model lacks a specific enzyme, mimicking the human disease. This mouse is used to study the potential use of liver stem cells to resolve the liver damage caused by the genetic disease. In addition, stem cells has discovered a liver stem cell line that can be cultured and developed into two different liver cell types; those of the bile duct and hepatocytes. Stem cells' work with pancreatic stem cells is chiefly centered on diabetes. Diabetes mellitus type 1 (DM1) is caused by a loss of function of the pancreatic beta islet cells.

Pancreatic beta islet cells produce and secrete insulin, a hormone critical for our metabolism and balance of blood sugar levels. Stem cell therapy could restore this cell population, potentially treating DM1. So far, monoclonal antibodies has been developed that can purify human pancreatic stem cells from a pancreatic tissue sample.

## STEM-LIKE CELLS, HUMAN BRAIN

There are different kinds of stem cells (e.g., embryonic, adult, and cancer stem cells). Stem cells are known for their innate ability to give rise to more developed daughter cells while at the same time maintaining a population of themselves. The asymmetric cellular divisions that give rise to different kinds of progenitor cells from stem cells, along with the unique environments in which they live, their so-called niches, are being studied in a variety of models with the goal of using these special cells and factors that control their growth and differentiation for tissue and organ repair. Embryonic stem cells can give rise to all tissues and organs, which in turn possess populations of so-called adult or somatic stem cells; even the adult human brain has an adult stem cell population that resides in a region around the fluid-filled spaces, or ventricles, which has the potential to be mobilized for brain and spinal cord repair. *In vitro* studies have replicated brain cell generation, or neurogenesis, from the post-natal and adult brain stem cells known as multipotent astrocytic stem cells (MASCs), isolated from biopsy and even postmortem human brain tissue specimens. MASCs are glial in origin, there are two major classes of cells in the nervous system, neuronal and glial cells, with the latter generally appreciated as being support cells for the message-sending neurons. An astrocyte, can behave as a stem-like cell in the adult human brain and can give rise to both neurons and glia (hence its multipotency). These MASCs are tapped as neuronal progenitors that can provide new neurons that are lost to injury or disease, including degenerative diseases such as Parkinson's disease and Alzheimer's disease, where neurons are lost, contributing to behavioural deficits. There is another class of glial cell in the adult human brain that exhibits neural progenitor cell behaviours, adult human neural progenitor cells, or AHNPs, which have been isolated and expanded from the adult human cerebral cortex. Using techniques similar to those developed for isolating, expanding, and characterizing embryonic stem cells and MASCs, novel cell culture conditions have revealed an unprecedented proliferative capacity of AHNPs such that a single proliferating AHNP, which again exhibits adult astrocyte qualities, can produce replacement cells for almost 50 million adult human brains. These AHNPs are target for drug-screening and pharmacological manipulation in the injured or diseased central nervous system with the potential for autologous or self-repair therapeutic approaches for neurological disease. There are strong similarities between the behaviours of adult brain MASCs, AHNPs, and astrocytic stem-like cells seen in human brain tumours, known as gliomas. Because stem and progenitor cells are so potent and plastic, their potential roles in tissue neo- and hyperplasia are important to study cancer stem cells that appear to be involved in the generation of very aggressive tumours including glioblastomas. The sources of cells that generate brain tumours could be MASCs or AHNPs that harbour genetic mutations that lead to their transformation and uncontrolled growth of cells. The ability of stem cells to seek out and attempt repair of compromised brain circuitries offers tremendous opportunities to harness their potency for the repair of neurological disease. The study and comparison of embryonic and adult brain stem-like cells during development and throughout life should provide insights into manipulating the normal and abnormal neurogenesis that occurs during brain development, regenerative attempts following injuries or diseases, and that seen in brain cancer.

## STROKE

Stroke is a medical condition in which brain cell death occurs due to a sudden reduction in blood flow to the brain. This reduction in blood flow diminishes oxygen delivery to the brain cells (neurons) and causes the cell deaths. The rupture of a blood vessel and blockage of a blood vessel are two common causes of stroke. Depending on where the damage is and the severity of the attack, stroke patients could suffer from a wide range of disabilities such as aphasia (a complete or partial inability to produce coherent speech), paralysis, and amnesia. Although the brain is capable of reproducing and repairing some of the cells lost or impaired during a stroke attack, this self-regenerating process still cannot fully offset the damage. However, with the advances in stem cell research, scientists are hopeful that certain neurons can be developed from embryonic stem cells to completely replace the ones lost during the attack, so that stroke patient might fully recover and regain the functions that were lost.

## TYPES OF STROKE

Strokes are identified according to how the attack occurs in the brain. Most strokes fall into two main categories: ischemic or hemorrhagic. There are two types of ischemic stroke; thrombotic and embolic. The blood clot in a thrombotic ischemic stroke is prompted by atherosclerosis, a condition in which the fat deposits in the artery. These fatty deposits harden into a thrombus (blood clot) inside the artery in the brain, it eventually prevents oxygen and nutrients from entering the brain and causes cell death. An embolic stoke is different in that the clot forms in the arteries or large blood vessels outside the brain. The clot often travels through the vascular system. An aneurysm, on the other hand, is responsible for a hemorrhagic stroke. An aneurysm is a balloon-shaped bulge that forms on an arterial wall. As blood enters and enlarges the bulge, the vessel wall can rupture. Consequently, the cells dependent on the nutrients carried by the blood cease to function and cell death occurs.

## SIGNS AND SYMPTOMS

Patients suffering from stroke may experience symptoms such as difficulty with walking, experiencing sudden numbness of certain body parts, and headaches. Moreover, patients could lose the ability to speak coherently and they could also suffer varying degrees of memory loss. Some patients could also suffer from sudden decreased vision or loss of coordination.

## DIAGNOSIS

A series of physical examination is done to determine and assess the patient's memory capacity, speech coherence, and sense of coordination. A computed tomography (CT) scan presents a clear view of the brain and the blood vessel abnormalities. Magnetic resonance imaging (MRI) provides a three-dimensional view of the brain which allows physicians to observe the extent of brain damage. Cerebral angiogram is another diagnostic tool to find the location of the clot. An X-ray image is taken after a dye is injected into blood vessels. The dye makes vessels more visible and enables physicians to determine the clot and the extent of the damage.

Physicians can administer drugs or perform surgeries on stroke patients. Drugs such as tissue plasminogen activator (tPA) dissolve blood clots and restore blood flow. Surgical options such as angioplasty and endarterectomy are also available to treat ischemic thrombotic strokes. In addition to drugs and surgeries, researchers have explored the use of stem cells to treat ischemic strokes. When stem cells are introduced into the damaged areas of the brain, these stem cells extracted from the bone marrow have the ability to differentiate into astrocytes and neurons, brain cells which are commonly damaged when strokes occur. Furthermore, these stem cells displayed the ability of migrating to the injured part of the brain. Investigators considered this study as a promising sign of the widespread use of stem cell therapy in treating stroke patients. Subsequent research into ischemic stroke treatment found that the mesenchymal stem cells (MSC), derived from the bone marrow, possess the ability to differentiate into neurons and other brain cells. Furthermore, the injected MSCs restore blood flow to the damaged part of the brain, hence nourishing the growth of neurons. Hemorrhagic stroke is treated differently with drugs and surgeries. Long-term rehabilitation is crucial to regaining some of the functions lost.

## TREATING STROKE IN RATS USING HUMAN EMBRYONIC STEM CELL-DERIVED NEURONS

After a stroke occurs, there is generally a loss of neurons (nerve cells) within the brain. Researchers hope to use embryonic stem cells to generate neurons to replace those lost cells. At present, they are able to drive human embryonic stem cells (hESCs) into becoming neurons. The transplants contain both desirable neurons and undesirable undifferentiated cells. Researchers are developing a cell culturing method that selects only human neural stem cells (hNSCs), which then drives them to become mature neurons with no undifferentiated cells remaining. The researchers hope to study these hESCs-derived neurons to learn how they differentiate and how they are different from human neurons. They hope to adapt this technique to treat human stroke patients.

# 7

# Batten's Disease

Batten's disease is a disorder with no cure. Batten's disease is a rare, fatal, autosomal recessive inherited disease affecting the nervous system (neurodegenerative disorder) that begins in childhood. It is named after the British pediatrician Fredrick Batten who first described the juvenile form of neuronal ceroid lipfuscinosis (NCL) in 1903. The name is now commonly used to encompass three forms of NCL: infantile, late infantile and juvenile onset. All forms have the same basic cause; lack of a lysosomal enzyme and have a similar progression and outcome. Batten's disease is caused when defective genes fail to make enzymes needed to dispose of waste made by brain cells. The waste piles up in the brain and kills healthy cells until the patient dies. Children with Battens' disease suffer seizures, progressive loss of motor skills, sight and mental capacity, eventually becoming blind, bedridden and unable to communicate. Today, Batten's disease is always fatal.

## Clinical Manifestations of Batten's Disease

Early symptoms of this disorder usually appear at age 4 to 10 years with a gradual onset of vision problems, including eye discoloration as a milky fog gloss over the eyes, or seizures. Early signs may be subtle personality and behaviour changes, slow learning or regression, repetitive speech or echolalia, clumsiness, or stumbling. There may be slow head growth in the infantile form, poor circulation in the lower extremities (legs and feet), decreased body fat and muscle mass, curvature of the spine, hyperventilation and/or breath-holding spells, grinding of the teeth, and constipation. Over time, affected children suffer mental impairment, worsening seizures, and progressive loss of sight, speech and motor skills. Eventually, children with Batten's disease become blind, bedridden, and demented. Batten's disease is often fatal by age 8 to 15.

## STEM CELL TREATMENT

A comprehensive program has been developed for patients suffering from Batten's disease by combining both nerve stem cell activation and stimulation treatment and stem cells injections. The treatment includes some medicines to adjust the microenvironment of the patient's brain, and in stem cell implantation, a "cocktail" of medications and neural nutrition materials are given through IV, biofeedback treatment and comprehensive rehabilitation. By employing all of the above methods we can maximally regulate and repair the patient's nervous system, improve symptoms such as motor disturbance and speech impediment. The curative effect depends on the patient's medical condition. In general, the earlier the patients receive the treatment, the better effect they can get. But this does not mean that we can cure the disease completely. Current treatment has only been proven to slow the progression of this terrible disease.

### Stem Cells injections

Before the injection of stem cells, medications will be used to improve the microenvironment of the brain. The disease is inherited in an autosomal recessive manner. The mutation causes the lack of lysosomal enzymes which can digest food particles and worn-out cell parts and it leads to the building-up of lipfusicins in the body's tissues. These substances consist of fats and proteins and form certain distinctive deposits that cause the symptoms. So at first doctors will give patients a medication to clean the waste in the brain and adjust the microenvironment of the brain. This pretreatment is very important as it can help stem cells make full advantage of the clinical effect. And with the cooperation of the medication and the stem cells treatment, the patients can get the greatest improvement.

## Neural Stem Cells injection

As an inherited disease with genetic defect, patients suffering from Batten's disease are unable to use their autologous stem cells to do the stem cells treatment. Hence neural stem cells treatment is the only remedy and is carried out by the implantation. Mesenchymal stem cells (MSC) are injected through the SCF (spinal cord fluid) by means of a lumbar puncture. Each injection has 1 million stem cells. The cells involved in the injection are safe to young patients because it has been calculated accurately beforehand. This procedure is short and simple, and it usually is done with local anesthesia. Yet, in order to prevent any risk of infection we maintain a strictly sterile environment. The procedure is done in the operating room by an experienced neurology doctor only. Neural stem cells are multipotent stem cells that can differentiate and develop into a variety of cell types, including neurons and neural ganglia cells. These new neurons and neural ganglia cells have normal functioning such as secreting neuro-factors essential to maintain the normal activity of the brain. The neural stem cells are acquired conveniently and all the neural stem cells come from Peking University's Stem Cells Research Center which is the largest and most qualified stem cell laboratory in China, where the cells have undergone rigorous tests according to our highly strict regulations. The advantages are: Firstly, they are convenient and timesaving because they do not have to be cultured for a long time. Secondly, the neural stem cells are safe and secure in clinical therapy. In addition, these stem cells have good biological activity and have little chance to create immune response as that of immunological rejection to the patients. The side effects of stem cells treatment are as follows: The procedure can cause minor side effect for 2 to 3 days after the stem cells injection, such as fever, sleepiness, headaches, dizziness and fatigue. These side effects are expected and should not alarm or concern you. Some patients do not suffer any side effects.

## Rehabilitation

Rehabilitation therapy is very much a necessary part of the entire treatment regimen. The stem cell mobilization technique combined with signal induction in the rehabilitation training procedure significantly improves the patient's neural functions. The rehabilitation methods are Physical

Therapy, Occupational Therapy, Speech Therapy and Traditional Chinese Medicine (TCM).

## TREATMENT PERIOD

The recommended period of treatment is 5–6 weeks which includes 3–4 stem cells injections.

## MEDICAL PLAN PROCEDURE

To start with, the waste/toxic substances accumulated in the brain are cleansed and the brain internal microenvironment is improved so that it is suitable for the subsistence of stem cells. The neural stem cells are implanted by subarachnoid cavity injection and the signal inductive treatment is carried out so as to fix the stem cell implants in the implanted location.

# 8

# Stem Cell Research and Kidney Disease

## BASIC BUILDING BLOCKS

Cells comprise the body's basic building blocks. They are found in the skin, muscles, bones, and the internal organs and are important indicators as to how our bodies function. There are perhaps thousands of specialized cell types in the adult human body. Their purpose is to perform specific functions for the organs or tissues which they comprise. Those cells that have shown the greatest promise of replacing diseased organs with healthy new ones are stem cells. As we know stem cells are unspecialized entities that distinguish themselves from other body cells in two important ways: (1) their numbers can be replenished for long periods of time by means of their division; and (2) once they receive certain chemical signals, they can transform themselves into specialized units with specific functions such as a heart or nerve cell.

## CELL TYPES

Stem cells can develop into different types, including:

- **Totipotent stem cells** can become any of the body's cell types, including the placenta, which nourishes the embryo.

- **Pluripotent stem cells** descend from the embryo's totipotent stem cells. They can differentiate into any type of cell except for totipotent and placenta.
- **Multipotent stem cells** These descendants of pluripotent stem cells are primarily found in the bone marrow and "give rise to all of the cells found in the blood".
- **Progenitor cells** (unipotent stem cells) produce only red blood cells.

## ADULT VS. EMBRYONIC STEM CELLS

For some time, research scientists have been conducting studies to see if the stem cells found in the adult body have the same ability and promise for development as those in embryos. They discovered that ASCs appear to be less versatile while ESCs have far greater potential for treating and/or curing a wide variety of serious ailments. ASCs may be able to develop into only a limited number of cell types. ESCs can divide indefinitely when placed in a culture dish while this may not apply to ASCs, thereby reducing their capacity to form new types of cells.

### Embryonic stem cells

ESCs are produced from four-to-five day-old embryos. At this stage, they are known as "blastocytes". Scientists create ESC cultures by transferring a blastocyte's cell mass into a culture dish. The cells are then removed and placed into fresh culture dishes. After being repeated many times, millions of ESCs are eventually produced. Blastocytes that are utilized for treating ESC lines are got from eggs that were fertilized in *in vitro* fertilization clinics but were never placed in a woman's uterus.

### Adult Stem Cells

ASCs are found in smaller numbers in most adult tissues. Their primary function is to maintain and repair their host tissues. One advantage of using a patient's own ASCs is that they can be expanded in culture dishes, differentiated into the desired cells and reintroduced into the patient. Using the patient's cells would guarantee that they would not be rejected by the immune system. Compared to ESCs, however, there is a disadvantage in that it is harder to expand the numbers of ASCs in cell culture.

## NEW ORGANS

Most scientists seem convinced that the work done in stem cell research is so promising that it is only a matter of time before widespread therapy derived from this research is routinely used. This includes treating kidney disease and/or growing new organs for transplant. One major challenge revolves around tissue rejection. As in organ transplants, the immune cells in the body will attack transplanted cells as "foreign". This would trigger an immune rejection that could cause failure of the transplant and even endanger the life and welfare of the patient. But pluripotent stem cells could well be a source of replacement cells and tissues to treat many diseases, including Parkinson's and Alzheimer's, spinal cord injuries, stroke, burns, diabetes, osteoarthritis, and rheumatoid arthritis. Widespread use of this therapy could also help ease the great shortage of organs available for transplantation. The National Kidney Foundation believes that the stem cells have the ability to adapt and regenerate into different types of cells and they "have the potential to replace tissues damaged by disease. It is hoped that such tissue engineering might someday help doctors eliminate the need for many transplants and the anti-rejection drugs used in transplantation". Cloning new organs may be one way stem cells can benefit those waiting for donor organs. This involves "reprogramming a cell by replacing its nucleus with that of another cell so it becomes the generic equivalent of the original". Known as a nuclear transfer, the process "raises both hope as well as ethical concerns regarding the possibility of cloning humans for organs".

## EARLY-STAGE RESEARCH

As scientific investigation goes, stem cell therapy research is still in its infancy. Scientists have been able to do experiments with ESCs only since 1998. This is when James Thompson, MD., at the University of Wisconsin, used a technique he developed to both isolate and grow the cells. Real research was slow to continue, however, since federal funds to support limited ESC research have only been available from the time when

President George W. Bush announced the decision to fund it in August 2001. The funding applies to research using only the 21 ESC lines existing at that time. Currently, because of the mandated limited research on ESCs, ASCs are the only kind of stem cells commonly used to treat human diseases. Actually, doctors have been transferring hematopoietic stem cells (HSCs) in bone marrow transplants for more than 40 years. In recent years, scientists have developed more advanced techniques of "harvesting" HSCs to treat leukemia, lymphoma, and several inherited blood disorders.

## TREATING KIDNEY DAMAGE

In 2001, scientists from the Imperial Cancer Research Fund and the Imperial College School of Medicine, London, England, "discovered that cells in bone marrow can be turned into tissue, which could help treat kidney damage caused by cancer or other diseases." They studied female kidneys transplanted into male patients with bone marrow transplants. They discovered that "the bone marrow cells had transformed into kidney tissue".

# 9

# Stem Cell in Eye Diseases

"Stem cell treatment restores vision" study shows damaged corneas may be regenerated with patient's stem cells. Italian researchers report that the stem cell procedure resulted in successful corneal transplantation in three-fourths of patients with blindness in one or both eyes, caused in most patients by chemical or thermal burns. Vision was at least partially restored in patients who did not have major damage to other parts of the affected eye.

## REGENERATION OF CORNEAS

The procedure involved extracting healthy stem cells from the limbus, which is located between the colored and white part of the eye. This procedure can also be done even when only a tiny portion of the limbus remained undamaged. Stem cells taken from the biopsied limbus tissue grew into healthy corneal tissue in a little over two weeks, healthy tissue then grafted onto the damaged eye. The procedure was successful, the damaged, opaque cornea became clear again and the eye looked normal. People with corneal damage from chemical and thermal burns often have symptoms including light sensitivity, itching and pain. These symptoms went away or were much less severe in the successfully treated patients. Following successful transplant, about half of the patients had further surgeries to improve visual acuity and most showed at least some improvement in vision. Some patients achieved normal vision with the stem cell grafting alone.

## STEM CELL TREATMENT COULD CURE BLINDNESS

A revolutionary stem cell treatment developed by British scientists restored sight in the blind. Blindness has been cured by replacing worn out and damaged retinal cells with stem cells. Experts described the research as "stunning" and said the transplants could transform the lives of hundreds of thousands of Britons who have lost their sight—with the first operations taking place within a decade. Among those who could benefit are sufferers of macular degeneration—the most common cause of blindness in the elderly—and those who have lost their sight as a complication of diabetes. The researchers, from University College London and London's Moorfields Eye Hospital, used stem cells—blank cells with the power to turn into different types of cells—to restore the vision in blind mice. The cells were injected into the back of the eye where they replaced damaged photoreceptors—tiny light-sensitive cells found in the retina and key to vision. The decay and death of these delicate cells is responsible for blindness. The success of the technique was the selection of stem cells which were slightly more mature than the others. These cells both turned into photoreceptors and successfully connected with the nerves leading to the brain. Increasing the number and type of cells transplanted could improve sight even further. Most of the photoreceptors transplanted were rods—the cells used in peripheral and night vision. Transfer of cones, would greatly improve the result. While stem cells are often obtained from embryos in the first days of life, it is thought the patient's own eye could hold a bank of cells suitable for transplant. Recent research has shown that a population of cells found on the margin of the adult retina have stem cell-like properties, in other words, they are capable of self-renewal. These could be harvested through minor surgery and grown in the laboratory before being re-implanted onto the retina. Using the patient's own cells would also avoid the possibility of 'foreign' cells being rejected by the body. However, they are hopeful of the first human photoreceptor transplants taking place within ten years. Most of the patients have eye disease that results in death of photoreceptors. This research is the first to show photoreceptor transplantation is feasible. The technique could also be adapted to create stem cells capable of repairing the brain cells destroyed in Parkinson's disease or nerve cells damaged in spinal injuries. Eye experts described that the cells transplanted into the retina of mice can integrate and can function as photoreceptors. As with any basic research, this is a stunning piece of research that may in the distant future may lead to transplants in humans to relieve blindness and transform the lives of those with macular degeneration.

## STEM CELL RESEARCH IN EYE DISEASES

***Wisconsin research team creates retinal cells from skin cells*** Researchers from University of Florida chemically induced bone marrow stem cells to become retinal cells. Now University of Wisconsin researchers have broadened the Florida result and began the project by using embryonic (ES) stem cells. Ultimately, multiple types of retinal cells are grown beginning with either ES or iPS cells, each of which provided a highly enriched population of very primitive cells with the potential to become retinal cells. In normal human development, embryonic stem cells begin to differentiate into more specialized cell types about five days after fertilization. The retina develops from a group of cells that arise during the earliest stages of the developing nervous system. The Wisconsin team used cells from skin, created induced pluripotent stem cells, and then coaxed them into becoming retinal cell types.

***Researchers Coax bone marrow stem cells to become vision cells*** On July $26^{th}$ 2009, some of the researchers working on stem cell solutions for eye diseases used bone marrow stem cells programmed to repair damaged retinas in mice. A virus carrying a gene was used to gently push the cultured adult stem cells from mice towards retinal cells. However, only after the stem cells were reintroduced into the mice they completely transform into the desired vision cells, apparently taking environmental cues from the damaged retinas. After studying the cell-transformation process, University of Florida researchers were able to bypass the gene manipulation step entirely and instead use chemical compounds that mirrored environmental conditions in the body, thus pointing the stem cells towards their ultimate identities as vision cells.

***Corneal transplant and reconstruction using stem cells*** Dr. Jurkunas is a researcher at the Harvard Medical School's Centre for Cornea Research. The cornea is the clear, hard part of the eye, through which light enters.

To function ideally, the cornea must be clear, smooth and well hydrated (lots of water). Most of the water is contained in the complex, protective tear film that overlies the cornea and the sclera (white part). The tear film is the main refractor of light into the eye and retina. Dr.Jurkunas stated that "good hydration equals good vision" and that the environment is very hazardous to the eyes and that the cornea is the most vulnerable part of the eye. With too little water in the tear film over the cornea, it becomes irritated, vision becomes blurry and corneal ulcers may form. This can happen to anyone but especially the elderly. Corneal stem cells are formed on the surface of the sclera and in the corneal epithelium (membrane over the cornea). They are critical to keeping the cornea healthy and clear but they can become depleted or impaired. Causes may be overuse of contact lenses, infection, allergies, chemical burns and immunological disorders. There is currently no treatment for this. Dr.Jurkunas' research involved removing stem cells from the mouth and conjunctiva (meaty corner of the eye), or from the good eye, growing them in culture and transplanting them to the affected cornea. Older people with severe vision loss due to corneal damage are the best candidates. It will not help macular degeneration.

***Regarding computers*** Excessive computer use tends to reduce blink rate and therefore effect tear film hydration. Best is to correct this before it becomes serious. He has also co-related with other environmental factors that affect the eyes (regarding cataracts and macular degeneration). Eat foods and nutrition supplements high in antioxidants. Also, protect your eyes from UV light. Children should wear wraparound sunglasses.

## STEM CELLS REPAIRING A DAMAGED OPTIC NERVE

The optic nerve is a cable of nerve fibres that carry electrical impulses, containing visual information, from the eye to the brain. In adult mammals, any damage to the optic nerve caused by injury or disease tends to be permanent, because the cells that form the optic nerve cannot regenerate or repair themselves. This is why glaucoma and other diseases that involve optic nerve damage, such as optic neuritis and optic nerve trauma, lead to permanent loss of vision. Similarly, nerve fibres in the brain and spinal cord are unable to regenerate. Regeneration is necessary for restoration of function following damage, through accident or disease, to the human brain and spinal cord. The success of optic nerve regeneration opens also the door to future strategies of eye transplantation. Scientists around the world are working hard to find a way to regrow the optic nerve. In addition, because of its easy access, the optic nerve has long served as a standard model for the study of spinal cord injury and regeneration in laboratory animals. Only recently has it become accepted that the failure of the optic nerve to regenerate is not a single-factor event. Successful restoration of vision after optic nerve injury can be divided into five necessary steps: (1) First, damaged neurons must be able to stay alive. (2) The surviving neurons have to be reprogrammed to turn on their growth machinery for nerve re-elongation. (3) The regrowing nerve fibres should be able to penetrate injury-generated scar tissue. (4) The regenerating fibres must overcome growth-inhibitory signals on the path they navigate. (5) Finally, the fibres also have to be guided back appropriately to their original target to re-form functional connections. Each of these steps presents a great challenge to our efforts to repair an injured optic nerve. In the past, most attempts to promote optic nerve regeneration focused on only one of these steps. Some success in optic nerve regeneration, albeit very limited, was achieved: For example, Albert Aguayo's (a Neuroscientist) group performed a study using growth factors, such as brain-derived nerve growth factor, a support neuron survival. They demonstrated a significant increase in the number of neurons surviving after injury; however, disappointingly, they found little improvement in optic nerve regeneration beyond the site of the injury. Monsul and Hoffman in the Neuro-ophthalmology Laboratory at Johns Hopkins University injected cyclic adenosine monophosphate (cyclic AMP) into the vitreous of mice to boost the neurons' intrinsic growth potential for nerve re-elongation. They found that cyclic AMP was effective in inducing a small number (less than 1 percent) of nerve fibres to regenerate, but only for a short distance (up to 1 mm), through the crush site. Lehman and colleagues used an enzyme called C3 to enhance neurons' ability to penetrate glial scars and, at the same time, block nerve-growth inhibitory signals. They found that about 0.5% of nerve fibres regenerated within the injured optic nerve. Similarly, in collaboration with a group at Children Hospital they showed that blocking EGF-receptor signaling suppressed neurons' responses to glial scars and to nerve-growth inhibitors. As a result, approximately 0.5% of severed optic nerve fibres regrew past the injury site and elongated up to 2 mm. Recently, an important breakthrough by using combined approaches proved that 40–70% of optic nerve fibers regenerated from the eye all the way into the brain (about 7 mm) in mice, at ages up to two weeks after birth; this is at a time when nerve-growth inhibitors have already appeared in the mouse brain. Further studies shows that the functional restoration of sight after optic nerve injury may become possible in the near future, first in mice and then in people. With recent progress in stem cell research, neural stem cell therapy is emerging as a new approach that may be promising for restoration of sight after optic nerve injury. Attempts have focused on activating dormant stem cells, already present in the eye, to participate in the repair process in response to injury. In addition, it may be possible to harvest and transplant donor stem cells into the eyes of patients whose optic nerves have been injured, to replace the neurons that have been lost or to provide a permissive environment for nerve regrowth. Although these studies are still in their infancy, the field is progressing rapidly and may soon turn the stem cell approach into a viable therapy.

## EYE DISEASES

Dystrophies make the second largest group of eye diseases after inflammations. By clinical classification, dystrophies can be primary (no past history of ophthalmological diseases) and secondary (caused by inflammations and injuries). Foetal stem cells offer gigantic potential for correction and inhibition/termination of dystrophy progression. Eye tissue quality improvement triggered by capillary network growth, and optical nerve quality improvement after stem cell therapy result in long-term remissions of different retinal disorders (enzymopathies affecting cell quality, nucleotide/chromoprotein metabolism disorders, damage of neuroreceptors, retinal pigment epithelium and retinal ganglion cells). Best's and Stargardt's diseases are hereditary primary pigmentary (tapetoretinal) dystrophies affecting mainly the macula (yellow spot). Hereditary central dystrophies and abiotrophies ("abiotrophy"—lack of vital capacity) develop at a certain age, affecting several members of the same family, causing identical damage in both eyes, and they are triggered by a genetic defect causing impairment of enzyme systems and metabolic processes associated with them. Stargardt disease is the most common form of inherited juvenile macular degeneration that begins in late childhood and leads to legal blindness. The mutations are in a gene called *ABCA*4 responsible for the majority of Stargardt disease in children and the FDA approved a Phase I/II multicenter clinical trial using retinal cells derived from human embryonic stem cells (hESCs) to treat patients with Stargardt's Macular Dystrophy (SMD).

## Stargardt's Disease

Stargardt's disease is an autosomal recessive hereditary macular degeneration described by K.Stargardt in 1909-1913. Stargardt's disease may occur in one of every 20,000 children over the age of 6 and is usually diagnosed before the age of 20. It gradually leads to marked vision loss. The process is localized in the macula and extends to the surrounding retina. Stargardt's disease is caused by heterozygosity of *ABCA*4 (used to be called *ABCR*) gene. In this disease, loss of central vision may be due to foveal cone degeneration caused by mutation of *ABCA*4 gene present in foveal and perifoveal cones. There are 3 stages of the disease:

**Stage I** Photophobia (dread of light) for no apparent reason, visual acuity reduction to 0.5–0.6.

***Opthalmoscopic findings*** Optic disc and retinal arteries are normal; macular reflexes either disappear or become blurred; macula is greyish or greyish-reddish; retinal edema in macular region. Visual field borders, scotopic (dark) adaptation and colour perception are normal.

**Stage II** Children complain of marked vision loss (0.1–0.2), photophobia.

***Opthalmoscopic findings*** Slight blurring of upper and lower borders of the optic disc due to oedema of nervous fibre layer; dilution of the veins. Macular reflexes are blurred and polymorphic. Defects in central visual field are not usually found because fixation point is "floating", colour sensitivity threshold increases, and scotopic adaptation remains unchanged. It is quite difficult to differentiate between stages II and III, and it usually takes around 6 months from stage II to progress to stage III.

**Stage III** Marked vision loss to 0.0.

***Ophthalmoscopic findings*** Macular reflexes are blurred or absent; retinal pigment epithelium dystrophy with dust-like pigment deposits and multiple small greyish, though sometimes large, irregular lesions with metallic reflection. Scotomas in the visual field. Colour sensitivity thresholds increase, mainly to red colour, red-green dyschromatopsia, scotopic adaptation remains unchanged, unless there is peripheral dystrophy. Stargardt's disease is usually treated with oxyribonucleic acid and taufonum derivatives.

## Best's Disease

Best's disease—Egg-yolk (vitelliform) macular dystrophy described by Best in 1905. First symptoms of the disease appear at different age from newborns to adults over 30. Best's is autosomal dominant. Presently, abnormal genes are identified on chromosome 11, but the gene causing this condition is not yet known (since recently, *VMD2* gene is associated with Best's disease).

Best's disease develops in stages and can be monolateral.

**Stage I** ("egg-yolk lesion")—Large cyst-like lesion with clear or yellowish content appears in the macula. This lesion has clear borders irradiating into the vitreous body. Both variants are possible within the same family; central scotoma is absent and normal or subnormal visual acuity.

**Stage II** ("scrambled egg")–Oedema and aggregation of exudatation clots in the lesion, haemorrhages are also possible. Gradually, oedema and haemorrhages disappear, and the cyst is substituted with flat atrophic yellowish focus "powdered" with fine pigment.

***Morphology*** Destruction of pigment ephithelium in the macula, atrophy of choriocapillaries, thinning of Bruch's membrane, partial neuroreceptors. Visual acuity reduces to 0.0, seldom to 0.1–0.2. Preservation of relatively high visual acuity happens when the lesion is located behind neural epithelium. Currently, there are no effective methods of Best's disease treatment. Biostimulators provide some stabilization. In case of haemorrhages and exudation, the patient is usually prescribed resolving, dehydration and nonspecific anti-dystrophy therapy. The patient should try to prevent focal infections.

## Congenital Blindness and Eye Diseases

Genetic diseases of the eye can arise from one of 160 mutations. Because the eye does not normally mount an immune response, the eye may not reject the new genetic material or the new cells as readily as other tissues. Gene therapy trials for 2 eye diseases, Leber congenital amaurosis (LCA) and Stargardt disease, are ongoing or recently completed. Transplantation of limbal epithelial stem cells for treatment of their deficiency is the first cell therapy for ocular diseases in clinical practice. Patients enrolled in clinical trials have observed benefits which provide proof of feasibility of treating damaged eyes. Development of gene and cell therapies for eye diseases is a hotly pursued field. LCA is a rare disease that can be caused by one of 15 genetically changed genes, such as the retinal protein RPE65 that is crucial for visual stimulation from the retina to the brain. Although the protein does not maintain its function, the retinal cells remain alive and wired into the optic nerve for sight. Twelve patients with defective RPE65 were treated by gene therapy to correct this genetic defect in a phase I clinical trial. They were administered a low, medium or high dose of a viral vector that could produce the normal RPE65 protein in the retina of afflicted patients. Scientists and clinicians are also interested in treating other hereditary and non-hereditary eye diseases. In some eye diseases, different cell populations, including the retinal cells, rods, cones or limbal epithelial stem cells atrophy or die. Thus, stem cells or stem cells expressing the corrected/normal gene would be needed because of the lack of intact cells. For example, limbal epithelial stem cells can also die by exposure to chemicals and their deficiency causes pain, light sensitivity, and cloudy vision. Transplantation of limbal epithelial stem cells for treatment of this deficiency is the first cell therapy for ocular diseases in clinical practice. These human studies provide proof of feasibility of replacing damaged or lost cells in the eye with stem cell therapy.

# 10

# Burns

Burns are injuries to tissues caused by heat, chemicals, radiations, friction or electricity. Burns injure the skin layers, and they may also injure other parts of the body such as muscles, nerve, lungs, and the eyes. Depending on the severity and depth of tissue damage, burns are classified into various degrees; first-degree burns are superficial thickness burns, extending only into the epidermis (outermost layer of the skin); second-degree burns, which are partial thickness burns, involve the superficial dermis (the layer of the skin deep to the epidermis, consisting of a bed of vascular connective tissue and containing the nerves and organs of sensation, the hair roots, and sebaceous, and sweat glands) and possibly also the deep dermis; third-degree burns are full-thickness burns that include the epidermis, the entire dermis, skin appendages, and at times deeper tissue; finally, in fourth-degree burns, tissue is destroyed to the level of or below the deep fascia. Traditionally, partial-thickness and full-thickness burns are treated by skin grafting. Skin grafting is a two-step surgical procedure in which a patch of skin is removed from one area of the body and transplanted to cover the wound. This procedure is effective but leaves the patient with a scar and may take months to heal. However, in cases of severe burns where there is not any healthy skin left to graft or the patient is not healthy enough to undergo such a surgical procedure, stem-cell based therapy provides a promising alternative.

## STEM CELL-BASED THERAPY

The basic concept of stem cell therapy in burns, and skin replacement in general, involves growing stem cells on a synthetic scaffold, and then transferring that medium on to a patient's wound or burn. The goal of this research area is to replace conventional skin grafts with a new method of healing that has better results, a faster healing rate, and no complications or rejection by the patient. Embryonic stem cells derived by somatic cell nuclear transfer are not rejected by the recipient, who is the donor of somatic cells. In this procedure, DNA from a somatic cell (a body cell other than a sperm or egg cell) is removed, and the rest of the cell is discarded. Then this DNA is inserted into an egg cell, the DNA of which has been removed. After being inserted into the host cell, this DNA is reprogrammed by the egg. Embryonic stem cells obtained in this way can be used to develop therapies that are ideally suited for a specific person. Embryonic stem cells can be induced to differentiate into skin keratinocytes in *in vitro* cultures under specific conditions and can be used for generating artificial skin—an unlimited supply of graftable regenerated epithelium that is without rejection risk. Skin has tremendous potential to regenerate after injuries. The epidermis has cells (keratinocytes) that resemble adult stem cells. These are classified on the basis of their capacities for multiplication before they undergo differentiation. These are holoclone keratinocytes (which have tremendous potential for differentiation and can undergo 140 divisions before senescence), paraclone keratinocytes (have limited growth potential, undergo a maximum of 15 divisions), and meroclone keratinocytes (which have a proliferative potential greater than paraclones). It is holoclone keratinocytes that can be cultured *in vitro* in the presence of epidermal growth factors to produce or supply stratified epithelium, which retains the properties of the original epidermis. Autologous (from the patients own body) cultured keratinocytes (i.e., holoclones) can be used to treat massive full-thickness burns. Holoclone keratinocytes grafted onto a burn site effectively proliferate and promote epidermal would healing. However, the actual usefulness of this procedure in treating deep full-thickness burns is limited by the fact that epidermal keratinocyte sheets grown by this method contain no dermal factors. This has led to investigation into the use of some kind of dermal element or substrate to support the keratinocyte layer.

The keratinocytes are usually grown on the chosen substrates. The substrates that have been used include collagen gels (collagen is the main protein of connective tissue in animals) and cryopreserved dermis (cryopreservation is a process in which cells or whole tissues are preserved by cooling to low subzero temperatures). Swiss doctors have also explored the potential of foetal skin cells in grafts. The therapy involves the use of skin cells grown from aborted foetuses to heal burns, sparing the need for skin grafts. Foetal cells are known to have remarkable regenerative potential. The foetal tissue promotes the growth of patients' own skin so it can be used as a biological bandage. The researchers collected skin from aborted foetuses and allowed the cells to divide *in vitro*. The cells were then allowed to grow on a bed of collagen (an important protein in skin). This procedure can be used to obtain several million—100 cm—patches for use in transplants from a single foetal biopsy (the removal of a sample of tissue from a living person for laboratory examination). The patches obtained in this manner were then placed on burn wounds, which took an average of 15 days to heal-considerably faster than traditional skin grafting (almost six times faster). This results in almost a perfect skin. There was little hypertrophy of the new skin and no retraction. None of the foetal cells remained in the healed skin. Grafts acts as growth factor and cytokine secretors rather than true grafts. Human umbilical cord blood stem cells differentiate into epithelial cells under specific *in vitro* conditions and therefore can also be used as a source of artificial skin for patients with deep burn injuries. Similarly, haemopoietic stem cells of the bone marrow can differentiate into blood cells. Studies have shown that these haemopoietic stem cells, when transplanted on deep burn wounds, decreased the healing time. Thus, haemopoietic stem cells cultivated *in vitro* may prove to be useful in burn treatment. Growing new skin is actually possible. In 2008, information based on clinical trials was released by Stratatech Corp., a Wisconsin biotech company. Their cell line was derived from keratinocytes, the most common and least immunogenic of the cells found in the epidermis. The first round of clinical trials used 15 patients, with problems ranging from burns to flesh-eating bacteria at the University of Wisconsin Hospital in Madison and at the Arizona Burn Center in Phoenix. The company's tissue was used to cover the wound for one to two weeks to prepare the wound for a transplant of the patient's own skin. In their second round of tests, the genetically altered patch remains on the patient, instead of removing the patches and replacing them with grafts of the patient's own skin. Stratatech Corp., is involved in developing three lines of genetically engineered material that would enhance their cell line with extra healing powers. A first version would stimulate the growth of new blood vessels. A second version would be antimicrobial to ward off infection. A third version would neutralize substances believed to be a factor in creating skin ulcers.

Stem cell for the treatment of spinal cord injury is still in its primary state and there are several promising projects that are preparing to enter the clinical study phase. Physiologically, spinal cord trauma differs from other types of central nervous system injuries. Instead of cell bodies (as in the brain), the spine is predominantly composed of long axon tracts. Spinal cord damage results in loss of sensory, motor, or autonomic function. More specifically, damage to the spinal cord prevents transmission of sensory information to the brain as well as transmission of motor and autonomic commands from the brain to the body. Effectively, areas below the level of injury lose sensation and control, which manifests as numbness and paralysis. The total loss of sensation and control is considered a "complete" spinal cord injury whereas partial loss of function is "incomplete". Characteristic of the central nervous system, neurons of the spinal cord have limited regenerative abilities, rendering most injuries permanent. The advantage of damaging an axon is that it does not necessarily imply the death of its corresponding neuron. Rather through a process known as Wallerian degeneration, only the severed portion of an axon is lost; the remaining, nucleus-containing segment remains viable, with the potential for regrowth. However, the inhibitory bio-chemical signals produced by the microenvironment of the spine seems to be partially responsible for its limited regenerative properties and an associated process of Wallerian degeneration is the demyelination of damaged axons. This important component of neurons normally surrounds the axon tract to effectively serve as an electrical insulator. Through this mechanism, speeds of signal propagation through the axon can increase up to 70-fold. Without myelination, axons in the spine transmit information too slowly to support a functional human being.

## CLINICAL TRIALS USING OLFACTORY ENSHEATHING CELLS

One of the FOR-SCI priorities was a 2002 study conducted by Jike Lu and colleagues. The experiment asserted the ability of olfactory ensheathing cells (OECs) to promote recovery of motor functions. Furthermore, the protocol used in this study was said to be effective even after a month long delay between injury and transplant. The success of the project was said to stem from the use of OECs. Functionally, olfactory neurons are unusual in that they have regenerative abilities because of their constant exposure to the external environment (much like how skin is continually renewed). As a supporting cell, OECs surround the growing axons of olfactory neurons and stimulate them toward their target. In experiments on animal models, biopsies of rat and human nasal cavity linings (olfactory lamina propria) were taken and transplanted into damaged rat spines. The experiment used the lamina propria to function doubly as a reservoir for OECs and as a neuron-guiding cellular scaffold within the damaged region. Within two to three weeks of the procedure, motor abilities of the test subjects had already surpassed those of their untreated control counterparts. This progressive improvement of the test rats continued for the remainder of the study, while the control group remained at the same deficient level. Thus it is obvious that there are benefits facilitating functional improvement. This study also provided solutions to many questions plaguing stem cell researchers across the board. First and foremost, nasal cavity lamina propria was shown to be not only a reliable source of OECs but also to exist in relatively high quantities and is easy to obtain. Thus, the often controversial issue of foetal tissue is easily bypassed, and a clinically practical source of therapeutic cells has been identified. Second, the use of differentiated supporting cells and not embryonic stem cells eliminates the possibility of uncontrolled differentiation, and the possible proliferation, of implanted cells. Third, because tissues are patient derived, there is no fear of immunologic rejection. Subsequently, the patient does not have to undergo dangerous, lifelong immunosuppression. Fourth, the experiment appears to have been unaffected despite a month long delay between injury and treatment. Physicians prefer to wait at least a week after spinal trauma before beginning any interventions, to allow the inflammatory process to resolve. The success of this study holds strong implications for the development of a clinically viable protocol.

## TRIALS OF THE NEAR FUTURE

### Oligodendrocyte Progenitor Cells

The oligodendrocyte progenitors are derived from human embryonic stem cells and these oligodendrocyte precursors are functionally comparable to OECs. Both cell types support neurons, myelinate axons, and secrete a host of growth factors. These cells were induced by a series of signaling chemicals to differentiate into oligodendrocyte progenitors and then subsequently encouraged to proliferate en masse. The result of this procedure created high purity samples of progenitor cells, and in practical quantities. From there, researchers hypothesized that if the appropriate stimuli were administered simultaneously with the tissue transplant, progenitors would further differentiate into mature oligodendrocytes, where they could then remyelinate damaged axons. The transplant procedure was the same as above with a minor difference in the time delay before initiation of therapy. Instead of waiting four weeks between injury and treatment, transplantation was timed to occur at either seven days or 10 months following an injury. Results of the the implantations study demonstrated functional improvements and increased remyelination of damaged axons by day seven. Hence the most significant finding of this series of experiments is the fact that the oligodendrocyte progenitors seem to have no deleterious effects on normal tissue. With all the hype surrounding stem cell research, it is important to keep in mind that none of these studies, preclinical or clinical, have demonstrated a full recovery. Even a radical attempt at forcing a cure by repeated injections of 50 million embryonic stem cells directly into the spinal cord (as performed by researchers in India) has failed to restore subjects to their preinjury state. Nonetheless, these attempts have made some significant improvements in the lives of patients. Still, it is amazing how much stem cell research has achieved. Future studies will undoubtedly reveal more incredible qualities of these cells and how they can be used to develop a clinically viable cure for patients with spinal cord injuries.

# 11

# Stem Cell Therapy for Mental Disability

## FRAGILE X SYNDROME

Fragile x syndrome (FXS) is the most common inherited cause of mental disability. Although the precise number of people with this disorder is unknown, the Centers for Disease Control and Prevention has estimated that 1 in 4,000 males and 1 in 6,000 females have FXS. FXS is caused by the absence or dysfunction of a protein called the Fragile X mental retardation protein (FMRP). FMRP is essential to the nervous system because it is necessary for proper nerve maturation and for forming efficient nerve connections in developing embryos. The absence of FMRP in FXS patients is caused by a change, or mutation, in the Fragile X Mental Retardation 1 gene, *FMR*1. The *FMR*1 gene is responsible for producing the FMRP protein at the correct time during embryonic development. Currently, no cure exists for FXS. However, research on stem cells from humans, mice, and flies has helped to understand how the disease develops and may lead to improved treatment options. The symptoms of FXS can vary widely, with boys more often affected than girls. FXS most commonly affects intelligence, learning, physical appearance, social interactions, speech, and the senses. The intellectual disabilities in FXS patients range from mild learning disabilities to severe mental retardation. The physical symptoms of FXS are often undetectable in children; however, some teens and adults with FXS have long ears, faces, and jaws. Some patients have connective tissue defects, which cause flat feet, extendable joints, balance problems, and non-life threatening heart murmurs. About 20 percent of FXS patients are prone to seizures. Furthermore, autism-like behaviours (i.e., flapping hands, self-biting, repetitive actions, and walking on the toes) are also frequently observed in FXS children. Children with FXS may also have behavioural challenges that cause them to be extremely anxious in new social situations. Attention deficit disorder (ADD)/attention deficit hyperactivity disorder (ADHD) as well as autism are both common in FXS patients. Boys with FXS usually have some speech language problems. Most begin using words later in childhood than their unaffected peers. Speech problems may range from enunciation, stuttering, omitting parts of words, and misunderstanding tone of voice. FXS patients may also have difficulty comprehending body language. Finally, FXS patients can be extremely sensitive to sensory stimulation (e.g., bright light, loud noises, or skin sensitivity). Some children with FXS avoid being touched and have trouble making eye contact. The symptoms of FXS can be reduced and/or eliminated with treatment. Since FXS affects so many aspects of personal development, many types of treatment are used. Through special educational programs FXS children can progress in school, learn independence, and become employed as adults. Heart murmurs can be monitored by a physician, seizures treated with medication, and balance problems addressed with physical therapy. Speech therapy can improve speech impediments and help patients communicate, while behavioural therapy can help children cope with stress induced by social situations and sensory sensitivity.

## Stem Cell Research

*FMR*1 mutations affect the early development of neurons during foetal formation by changes in the gene works of the pre-nerve cells of a foetus. These pre-nerve cells are called neuronal stem cells or embryonic stem cells. The study of the embryonic stem cells derived from mice, flies, and humans show how the FMR1 gene is important for nerve maturation. Furthermore, studies have found that a decrease in FMRP protein ultimately alters nerve cell signalling and may affect the work of the nerves in the brain to learn and create memories. Drugs could possibly block these aberrant signalling events and improve the learning capability of FXS patients. Stem cells derived from animals have limited research potential because they lack the repeat sequences at the beginning of the *FMR*1 gene. Recently, scientists have developed human embryonic stem cells with the *FMR*1 gene defect. Studies of these cells have shown that the *FMR*1 gene is still functional in the earliest stages of the embryo and is shut off later during development. By understanding how and why the *FMR*1 gene gets turned off, scientist may be able to reverse this event in embryos with the FXS mutation.

## MULTIPLE SCLEROSIS

Multiple sclerosis (MS) is an inflammatory demyelinating, autoimmune disease of the central nervous system, with permanent lesion leading to disability. This disease makes the patient totally bedbound, resulting in bed sores and all kinds of infections to the disabled, depressed patient. Intensive therapy has a key role in managing this complex disease, and the slowing accumulation of neurological damage in MS has remained mostly irreversible. The therapies that improve blood circulation, mitochondrial energy production, and anti-inflammatory, antimicrobial, and antioxidant factors slows and perhaps reverses the disease with a comprehensive program that includes stem cell therapies. There are different forms of MS, including relapsing, remitting (attacks are followed by improvement), primary progressive (the disease progresses without remission), secondary progressive (initial remission followed by progressive disability), and relapsing progressive (disability occurs and accumulates between and during attacks). Although the cause of MS has not yet been found, neuro-researchers agree that whatever the cause, inflammation destroys the myelin sheath of the brain and spinal cord. The myelin sheath is an insulating structure around the nerves, which are responsible for the quick travelling of impulses. Once this myelin is affected, signal transmission is slowed down considerably, giving rise to the signs and symptoms of MS. In the brain and spinal cord, special glial cells, called oligodendrocytes, are responsible for the manufacture and maintenance of this myelin sheath. This is where the role of stem cells is being evaluated. When stem cells were injected into the demyelinated area of the brain and spinal cord, regrowth of the affected neurons is observed. Stem cells, both embryonic and adult, can potentially be used to deliver molecules that repair or help in the reformation of damaged tissue of the central nervous system. The disease is more common in females than males, but the disability is more pronounced when it affects males. A variety of signs and symptoms have been highlighted, including memory problems, decline in reasoning function of the brain, loss of coordination of body parts, paralysis, body weakness, urinary retention, urinary incontinence, sexual problems, visual problems, facial weakness, depression, pain symptoms, emotional disturbances, etc.

## Stem Cell Research

Because most of the active study involves mice, before human beings are involved, the first task for scientists was to find a lesion similar to MS in mice. Experimental autoimmune encephalomyelitis (EAE) was developed and found to be characteristically identical to MS in humans. The researchers then labelled neural stem cells of adult mice with iron particles and injected these labelled stem cells into the tail vein of mice with EAE. Labelling the neural stem cells with iron particles is of significance because iron particles interfere with magnetic fields and can be detected on magnetic resonance imaging (MRI). To the great satisfaction of the researchers, the injected neural stem cells started reaching the EAE—the damaged area—within the first 24 hours. Gradual improvement in symptoms appeared as the myelin sheath started to be repaired by the newly arrived stem cells. Many of the researchers used the embryonic stem cells for a successful treatment for patients with MS. One such study used embryonic stem cell suspensions containing stem cells obtained from active growing cells of four-to-eight-week-old cadaver embryos. MS patients were administered with these embryonic stem cell suspensions and were observed. This treatment resulted in an improved range and quality of motions in the extremities, normalized muscle tone, decreased fatigue and general weakness, and improved quality of life. Most of the patients involved in the study showed no further progression of disability. A purified, multipotent stem cell has now been isolated from bone marrow and umbilical cord blood; these stem cells have been named CD 34 (+). The CD 34(+) cell is capable of migrating to an injured or damaged area and proliferating into the specific cells required for the repair of that tissue. These cells basically transform into the neuroprotective glial cells, predominantly the oligodendrocytes and their progenitors. This potency of CD 34 (+) cells is now being applied for treating MS. Similar multipotent cells, called mesenchymal stem cells, which are found in the bone marrow and umbilical cord blood, have been extracted. These cells specialize into adipocytes, osteocytes, chondrocytes, or glial cells depending on the specific growth factors influencing them. Recently, a new procedure involving a patient's own bone marrow cells being injected, instead of embryo-derived cells, has been developed. This trial involves bone marrow stem cells of the same patient extracted from his or her bones, preferably the pelvic bones, and then being processed in the laboratory. The marrow stem cells are then effectively injected into the veins of the same MS patient. An MRI scan done before is then compared with the postinjection stage, for structural changes. These studies are in their initial stages, and a proven outcome in human beings is expected in the coming few years. Many researchers have been studying and applying the concepts of remyelination, which involves extraction of the stem cells and then culturing them in the laboratory. In this way, bone stem cells have been induced to differentiate into oligodendrocyte-like cells. This demonstration is of real hope for MS patients. Researchers have also found a factor termed Fyn tyrosine kinase, which can increase the production of myelin base protein. This myelin base protein plays an essential role in the remyelination of axons. The gene that codes for Fyn tyrosine kinase has also been isolated and termed the *fyn* gene. Another angle of focus is to first analyse and diagnose the deficient factor or gene in the respective MS patient. Once the defect is identified, a number of combinations can be tried, depending on the deficiency. For instance, for an individual who has a deficiency of fyn, incorporating umbilical cord stem cells containing the fyn gene will be of great benefit. All these studies have mostly been carried out on animal models, so that exact reaction of human beings is not perfectly understood. Moreover, scientists work out certain methods to use stem cells to help in the discovery of new genes and molecules that can cease the progression or help repair the damaged zone. These stem cells can also be of help in manufacturing new drugs that can treat MS. The use of embryonic and foetal stem cells for research purposes has been a cause of protest from different circles. Therefore, not all countries have been able to contribute to their full potential. The alternative methods are using stem cells from cadaver embryos, stem cells from umbilical cord blood, and recently, stem cells from the patient's own bone marrow, have opened new avenues for scientists and researchers. Many of the patients suffering from MS are now optimistic and improving and living an independent life because of this stem cell revolution. However, the work is still in its initial stages; with the hopeful outcome seen in mice models. Thus the future of stem cell therapy in treating patients of MS looks bright.

# 12

# Bone Marrow Transplantation and Peripheral Blood Stem Cell Transplantation

Bone marrow is the soft, sponge-like material found inside bones. It contains immature cells known as hematopoietic or blood-forming stem cells. (Hematopoietic stem cells are different from embryonic stem cells. Embryonic stem cells can develop into every type of cell in the body). Hematopoietic stem cells divide to form more blood-forming stem cells, or they mature into one of three types of blood cells: white blood cells, which fight infection; red blood cells, which carry oxygen; and platelets, which help the blood to clot. Most hematopoietic stem cells are found in the bone marrow, but some cells, called peripheral blood stem cells (PBSCs), are found in the bloodstream. Blood in the umbilical cord also contains hematopoietic stem cells. Cells from any of these sources can be used in transplants. Bone marrow transplantation (BMT) and peripheral blood stem cell transplantation (PBSCT) are procedures that restore stem cells that have been destroyed by high doses of chemotherapy and/or radiation therapy. There are three types of transplants:

- In autologous transplants, patients receive their own stem cells.

- In syngeneic transplants, patients receive stem cells from their identical twin.
- In allogeneic transplants, patients receive stem cells from their brother, sister, or parent. A person who is not related to the patient (an unrelated donor) also may be used. BMT and PBSCT are used in cancer treatment is to make it possible for patients to receive very high doses of chemotherapy and/ or radiation therapy. To understand more about why BMT and PBSCT are used. Chemotherapy and radiation therapy generally affect cells that divide rapidly. They are used to treat cancer because cancer cells divide more often than most healthy cells. However, because bone marrow cells also divide frequently, high-dose treatments can severely damage or destroy the patient's bone marrow. Without healthy bone marrow, the patient is no longer able to make the blood cells needed to carry oxygen, fight infection and prevent bleeding. BMT and PBSCT replace stem cells destroyed by treatment. The healthy, transplanted stem cells can restore the bone marrow's ability to produce the blood cells the patient needs. In some types of leukaemia, the graft-versus-tumour (GVT) effect that occurs after allogeneic BMT and PBSCT is crucial to the effectiveness of the treatment. GVT occurs when white blood cells from the donor (the graft) identify the cancer cells that remain in the patient's body after the chemotherapy and/or radiation therapy (the tumour) as foreign and attack them. To minimize potential side effects, doctors most often use transplanted stem cells that match the patient's own stem cells as closely as possible. People have different sets of proteins, called human leukocyte-associated (HLA) antigens, on the surface of their cells. The set of proteins, called the HLA type, is identified by a special blood test. In most cases, the success of allogeneic transplantation depends in part on how well the HLA antigens of the donor's stem cells match those of the recipient's stem cells. The higher the number of matching HLA antigens, the greater

the chance that the patient's body will accept the donor's stem cells. In general patients are less likely to develop a complication known as graft-versus-host disease (GVHD) if the stem cells of the donor and patient are closely matched. Close relatives, especially brothers and sisters, are more likely than unrelated people to be HLA-matched. However, only 25 to 35 percent of patients have an HLA-matched sibling. The chances of obtaining HLA-matched stem cells from an unrelated donor are slightly better, approximately 50 percent. Among unrelated donors, HLA-matching is greatly improved when the donor and recipient have the same ethnic and racial background. Although the number of donors is increasing overall, individuals from certain ethnic and racial groups still have a lower chance of finding a matching donor. Because identical twins have the same genes, they have the same set of HLA antigens. As a result, the patient's body will accept a transplant from an identical twin. However, identical twins represent a small number of all births, so syngeneic transplantation is rare. The stem cells used in BMT come from the liquid centre of the bone, called the marrow. In general, the procedure for obtaining bone marrow, which is called "harvesting", is similar for all three types of BMTs (autologous, syngeneic, and allogeneic). The donor is given either general anaesthesia, which puts the person to sleep during the procedure, or regional anaesthesia, which causes loss of feeling below the waist. Needles are inserted through the skin over the pelvic (hip) bone or, in rare cases, the sternum (breastbone), and into the bone marrow to draw the marrow out of the bone. Harvesting the marrow takes about an hour. The harvested bone marrow is then processed to remove blood and bone fragments. Harvested bone marrow can be combined with a preservative and frozen to keep the stem cells alive until they are needed. This technique is known as cryopreservation. Stem cells can be cryopreserved for many years. The stem cells used in PBSCT come from the bloodstream. A process called apheresis or leukapheresis is used to obtain PBSCs for transplantation. For 4 or 5 days before apheresis, the donor may be given a medication to increase the number of stem cells released into the blood stream. In apheresis, blood is removed through a large vein in the arm or a central venous catheter (a flexible tube that is placed in a large vein in the neck, chest or groin area). The blood goes through a machine that removes the stem cells. The blood is then returned to the donor and the collected cells are stored. Apheresis typically takes 4 to 6 hours. The stem cells are then frozen until they are given to the recipient. Stem cells may also be retrieved from umbilical cord blood. For this to occur, the mother must contact a cord blood bank before the baby's birth. The cord blood bank may request that she complete a questionnaire and give a small blood sample. Cord blood banks may be public or commercial. Public cord blood banks accept donations of cord blood and may provide the donated stem cells to another matched individual in their network. In contrast, commercial cord blood banks will store the cord blood for the family, in case it is needed later for the child or another family member. After the baby is born and the umbilical cord has been cut, blood is retrieved from the umbilical cord and placenta. This process poses minimal health risk to the mother or the child. If the mother agrees, the umbilical cord blood is processed and frozen for storage by the cord blood bank. Only a small amount of blood can be retrieved from the umbilical cord and placenta. So the collected stem cells are typically used for children or small adults. Because only a small amount of bone marrow is removed, donating usually does not pose any significant problems for the donor. The most serious risk associated with donating bone marrow involves the use of anaesthesia during the procedure. The area where the bone marrow was taken out may feel stiff or sore for a few days, and the donor may feel tired. Within a few weeks, the donor's body replaces the donated marrow; however, the time required for a donor to recover varies. Some people are back to their usual routine within 2 or 3 days, while others may take up to 3 to 4 weeks to fully recover their strength. Apheresis usually causes minimal discomfort. During apheresis, the person may feel light-headedness, chills, numbness around the lips, and cramping in the hands. Unlike bone marrow donation, PBSC donation does not require anaesthesia. The medication that is given to stimulate the mobilization (release) of stem cells from the marrow into the bloodstream may cause bone and muscle aches, headaches, fatigue, nausea, vomiting, and/or difficulty in sleeping. These side effects generally stop within 2 to 3 days of the last dose of the medication. After being treated with high-dose anticancer drugs and/or radiation, the patient receives the stem cells through an intravenous (IV) line just like a blood transfusion. This part of the transplant takes 1 to 5 hours. The stem cells used for autologous transplantation must be relatively free of cancer cells. The harvested cells can sometimes be treated before transplantation in a process known as "purging" to get rid of cancer cells. This process can remove some cancer cells from the harvested cells and minimize the chance that cancer will come back. Because purging may damage some healthy stem cells, more cells are obtained from the patient before the transplant so that enough healthy stem cells will remain after purging. After entering the bloodstream, the stem cells travel to the bone marrow, where they begin to produce new white blood cells, red blood cells, and platelets in a process known as "engraftment". Engraftment usually occurs within about 2 to 4 weeks after transplantation. Doctors monitor it by checking blood counts on a frequent basis. Complete recovery of immune function takes much longer, however up to several months for autologous transplant recipients and 1 to 2 years for patients receiving allogeneic or syngeneic transplants. Doctors evaluate the results of various blood tests to confirm that new blood cells are being produced and that the cancer has not returned. Bone marrow aspiration (the removal of a small sample of bone marrow through a needle for examination under a microscope) can also help doctors determine how well the new marrow is working. The major risk of both treatments is an increased susceptibility to infection and bleeding as a result of the high-dose cancer treatment. The antibiotics are used to prevent or treat infection. Transfusions of platelets and red blood cells are given to prevent bleeding and to treat anaemia respectively. Patients who undergo BMT and PBSCT may experience short-term side effects such as nausea, vomiting, fatigue, loss of appetite, mouth sores, hair loss, and skin reactions. Potential long-term risks include complications of the pre-transplant chemotherapy and radiation therapy such as infertility (the inability to produce children); cataracts (clouding of the lens of the eye, which causes loss of vision); secondary (new) cancers; and damage to the liver, kidneys, lungs, and/or heart. With alloegeneic transplants, GVHD sometimes develops when white blood cells from the donor (the graft) identify cells in the patient's body (the host) as foreign and attack them. The most commonly damaged organs are the skin, liver, and intestines. This complication can develop within a few weeks of the transplant (acute GVHD) or much later (chronic GVHD). To prevent this complication, the patient may receive medications that suppress the immune system. Additionally, the donated stem cells can be treated to remove the white blood cells that cause GVHD in a process called "T-cell depletion". If GVHD develops, it can be very serious and is treated with steroids or other immunosuppressive agents. GVHD can be difficult to treat, but some studies suggest that patients with leukaemia who develop GVHD have less chances of cancer to recur. Clinical trials are being conducted to find ways to prevent and treat GVHD. A "mini-transplant" (also called a non-myeloablative or reduced-intensity transplant) is a type of allogeneic transplant. This approach is being studied in clinical trials for the treatment of several types of cancer, including leukaemia, lymphoma, multiple myeloma, and other cancers of the blood. A mini-transplant uses lower, less toxic doses of chemotherapy and/or radiation to prepare the patient for an allogeneic transplant. The use of lower doses of anticancer drugs and radiation eliminates some, but not all, of the patient's bone marrow. It also reduces the number of cancer cells and suppresses the patient's immune system to prevent rejection of the transplant. Unlike traditional BMT or PBSCT, cells from both the donor and the patient may exist in the patient's body for sometime after a mini-transplant. Once the cells from the donor begin to engraft, they may cause the GVT effect and work to destroy the cancer cells that were not eliminated by the anticancer drugs and/or radiation. To boost the GVT effect, the patient may be given an injection of the donor's white blood cells. This procedure is called a "donor lymphocyte infusion". Advances in treatment methods, including the use of PBSCT, have reduced the amount of time many patients must spend in the hospital by speeding recovery. This shorter recovery time has brought about a reduction in cost. However, because BMT and PBSCT are complicated technical procedures, they are very expensive. Many health insurance companies cover some of the costs of transplantation for certain types of cancer. Insurers may also cover a portion of the costs if special care is required when the patient returns home. There are options for relieving the financial burden associated with BMT and PBSCT. A hospital social worker is a valuable resource in planning for these financial needs. Federal Government programs and local service organizations may also be able to help. The National Cancer Institute's (NCI) Cancer Information Service (CIS) can provide patients and their families with additional information about sources of financial assistance. NCI is part of the National Institutes of Health.

# 13

# Clinical Trials within U.S. Heart Disease

Cardiovascular disease is the number one cause of death in the United States. Cell therapy is being investigated at a rapid pace worldwide and may revolutionize contemporary cardiovascular disease treatment. Stem cells can differentiate into nearly any mature cell phenotype and can self-renew indefinitely. Adult-derived "progenitor" cells have limited differentiation capability and cannot self-renew; however, these cells are more readily accessible. Bone marrow and peripheral blood circulating progenitor cells have been the focus of current human trials. Initial emphasis on cell transfer to replace myocardium has recently shifted to paracrine modulation of myocardial remodelling, mechanical strengthening of scar tissue, and promotion of tissue survival. Human trials have demonstrated feasibility of a variety of cell-based approaches with modest short-term benefits. Cells from autologous (one's own self) and allogeneic (from donors) sources have been transplanted in patients with heart disease, using a variety of delivery approaches. Three clinical syndromes are principal targets for cell therapy approaches; myocardial infarction (MI), chronic myocardial ischemia, and cardiomyopathy. Cell therapy attempts to prevent heart enlargement following MI, develop new blood vessels and increase blood flow for chronic myocardial ischemia, and regenerate contractile heart muscle for cardiomyopathy causing heart failure. Mechanisms underlying the benefits of progenitor cells observed in animal studies remain poorly understood. Transplanted cells may encourage the release of factors that locally activate adjacent native cells with beneficial effects. A few investigators suggest that cells "fuse" with native cells and thereby alter their behaviour in some beneficial way. Others suggest that transplanted progenitor cells transdifferentiate and that the daughter cells become the cells or blood vessels. Fusion and differentiation events are perhaps too rare to have meaningful treatment effect. Several adult progenitor cell types have been studied, and many are identified by cell surface antigens measured by flow cytometry. CD34+ cells and CD133+ cells identify endothelial progenitors and may be isolated from bone marrow, adipose tissue, peripheral blood, and umbilical cord. These cells may promote new blood vessel growth in areas of ischemia. Mesenchymal stromal cells (MSCs) may induce a therapeutic effect via paracrine mechanisms. Skeletal myoblasts represent an abundant autologous source of cells. Resident cardiac stem cells have been discovered to reside in protected niches in the heart. Early phase clinical trials using skeletal myoblast and resident cardiac stem cell transfer following *ex vivo* expansion are ongoing.

## MYOCARDIAL INFARCTION

MI occurs with a mortality of approximately 25 percent over three years. A proportion of these patients will develop adverse remodelling, leading to congestive heart failure and sudden death. Cellular therapy offers the potential to improve these outcomes. Strauer *et al* first reported therapeutic cell transfer of unfractionated bone marrow mononuclear cells in 10 patients. The method used was to perform a large volume bone marrow aspirate and then reinfuse the cells directly into the infarct-related coronary artery, using the catheters that are familiar to interventional cardiologists. Bone marrow mononuclear cells were aspirated five to nine days following emergent coronary stenting for acute MI. After overnight storage, cells were infused into the open infarct-related artery via a coronary balloon catheter. The balloon was inflated during infusion to prevent backflow of cells. At three months, infarct size decreased, wall movement increased, and cardiac perfusion was improved compared with a parallel control group. Several major trials in cellular therapy for recent MI patients have been performed.

The Transplantation of Progenitor Cells And Regeneration Enhancement in Acute Myocardial Infarction (TOPCARE-AMI) trial investigates both safety, feasibility, and potential effects on parameters of myocardial function of intracoronary infusion of either circulating progenitor cells (CPC) or bone marrow-derived progenitor cells (BMC) in patients with acute myocardial infarction (AMI). At six months, cardiac ejection fraction improved in the BM-treated patients. At four months, the infarct size was reduced; however, there was no significant improvement of ejection fraction, myocardial flow, or metabolism in Infarcted segments, determined using highly sensitive imaging techniques such as magnetic resonance imaging. The Autologous Stem Cell Transplantation in Acute MI trial was a double-blind, randomized trial of intracoronary BMNC infusion with 50 patients in treated and control groups. No improvement in infarct area, size or function with intracoronary BM cell therapy was observed at six months. The BMNC-treated group had a significant increase in left ventricular ejection fraction at four months, assessed by cine- ventriculography. Furthermore, the treatment group had a reduction in major cardiovascular adverse events at one year. The Myocardial Regeneration and Angiogenesis in MI with GCSF (granulocyte colony-stimulating factor) and Intracoronary Stem Cell Infusion mobilized the peripheral blood mononuclear cell infusion. The treated group showed improved myocardial function and perfusion. Cytokine therapy with GCSF is effective for mobilization of BM-derived progenitors into the peripheral circulation. Early clinical studies suggested that GCSF therapy after MI reduces ventricular remodelling and improves ejection fraction. Investigators are now focusing on transplanting bone marrow cell subpopulations, in part because of the modest performance of unfractionated cell products, especially endothelial progenitor cells, MSCs, and cardiac stem cells. Bone marrow-derived CD133+ endothelial progenitor cells when administered through an intracoronary route in post MI cases showed better prognosis. There was a significant increase in ejection fraction and fractional shortening with ventriculography.

## CHRONIC MYOCARDIAL ISCHEMIA

Clinical cell therapy information for chronic myocardial ischemia stems from small-scale, nonrandomized, uncontrolled trials. A few trials have tested direct injection of endothelial progenitor cells during concomitant surgical revascularization. Catheter-based endomyocardial delivery of bone marrow-derived cells has also been tested. These small-scale studies have demonstrated reduced angina frequency, improved functional capacity, and reduced ischemic burden with cell therapy. Losordo *et al.* demonstrated the feasibility, safety, and bioactivity of autologous CD34+ endothelial progenitor cell transplantation. Cells were harvested by apheresis following cytokine mobilization. Unfractionated mononuclear cells were enriched for CD34+. Cells were readministered into the cardiac muscle via a catheter-needle device guided by an electro-anatomic mapping system. Ischemic areas determined by single photon emission computed tomography (SPECT) were targets for cell delivery. There was a significant reduction in angina frequency at three months. Favourable trends were seen for nitro-glycerine use, exercise tolerance, angina classification, SPECT ischemia scores, and quality-of-life measures. Stamm *et al.* directly injected autologous, CD133+ enriched bone marrow-derived progenitor cells during coronary artery bypass surgery in patients with chronic myocardial ischemia. Modest improvements in global ventricular function and regional perfusion were observed. The Prospective Randomized Trial of Direct Endomyocardial Implantation of Bone Marrow Cells for Therapeutic Angiogenesis in Coronary Artery Diseases. Around 28 chronic ischemia patients showed catheter-directed intramyocardial improvements in exercise time, cardiac function, and perfusion. Stem cells for Chronic Myocardial Ischemia trial is a phase 2b, multicenter, randomized controlled, double-blind clinical trial in the United States. The aim of this trial is to test whether percutaneous endomyocardial injection of autologous CD34+ cells enriched from GCSF mobilized peripheral apheresis product can reduce angina symptoms in subjects with refractory symptomatic chronic myocardial ischemia.

## CONGESTIVE HEART FAILURE

Congestive heart failure (CHF) is a clinical syndrome resulting from myocardial injury that impairs the heart's ability to circulate blood sufficiently to meet the metabolic needs of the body. CHF is common and can lead to frequent hospitalization and sudden death. Initially, the principal goal of cell therapy for CHF was myocyte replacement. Newer concepts such as paracrine factor-induced reverse geometric remodelling and neoangiogenesis are being proposed as potential beneficial mechanisms for therapy. Autologous bone marrow mononuclear cells and skeletal myoblasts have been delivered in a small series of CHF patients with reduced ejection fractions by catheter-directed transendocardial injections, intracoronary infusion, and direct epicardial injections during concomitant coronary artery bypass surgery. Intramyocardial deposits of skeletal myoblasts lead to apparent ventricular arrhythmia, perhaps as a result of poor electrical integration within the myocardium. Nevertheless, short-term improvements in cardiac function combined with modest symptom relief was observed, forming the basis for future, larger-scale clinical trials. Cell therapy has the potential to revolutionize the management of cardiovascular disease. Larger-scale trials are underway to demonstrate the efficacy of cell therapy in patients with MI, chronic myocardial ischemia, and cardiomyopathy.

## MAMMARY STEM CELLS

The mammary gland is a dynamic organ, capable of extensive growth, regeneration, and remodelling. The growth demands of mammary gland during embryonic development and throughout adulthood, suggests a function for mammary stem/progenitor cells. In addition, testing of mammary stem cell activity in rodent models has revealed the presence of sub-populations of cells with the ability to recapitulate adult mammary gland development. Although the contribution of mammary stem cells to breast cancer has yet to be determined, it has been proposed that these cells may play a significant role in mammary tumorigenesis. Mammary glands are secretory organs derived from the epidermis, which are unique to all members of the class Mammalia. Mammary glands exist in both female and male mammal species, in males the mammary glands are rudimentary and nonfunctional and in females, the mammary gland consists of a branching network of ducts embedded in a fat pad, which coalesce at the nipple and respond to local and humoral factors to produce milk in response to suckling.

### Growth Demands for Mammary Epithelial Cells

There are four processes that demand cell growth and may be mediated by embryonic or adult mammary stem cells (1) specification and invagination of the embryonic mammary placode. Embryonic specification of mammary placodes starts along two epithelial ridges, known as milk lines that are formed in the epidermis. Expansion of the ductal tree during colonization of the mammary fat pads in juveniles. Ductal outgrowth occurs in response to the female endocrine environment, and is led by specialized, proliferative centres called terminal end buds. Ducts comprise heterogeneous cells, broadly divided into a basal layer of myoepithelial cells that produce and adhere to the basement membrane. Luminal cells can have multiple layers in actively dividing juvenile tissues and preneoplastic conditions.

## Cycle-associated Proliferation

The hormonal cycles typical of mammalian female can generate mitotic indices in mammary epithelia that are typical of pregnancy, but only briefly (hours long), and is followed by regression and cell apoptosis. Expansion and differentiation during pregnancy, to produce the lactation-competent mammary gland.

# STEM CELL THERAPY

Stem cell therapy for cardiovascular disorders

## Clinical Indications

Coronary artery disease

Myocardial Infarction

Congestive Heart Failure

Primary Pulmonary Hypertension

Dilated Cardiomyopathy

## Role of Stemcure

Blood/Bone marrow sample of the patient is collected aseptically at hospital. On receiving the sample from the hospital we process them. Stem cells were isolated using specialized techniques and cultured for 5–6 days in controlled tissue culture environment using specialized media, growth factor and human skill into EPC (Endotheliel progenitor Cell). The processed stem cells are supplied to clinician for injection into the coronary arteries (intra coronary)/directly into the heart muscles (intra myocardium)/through intravenous depending on the prevailing condition.

## Results

Recent scientific studies show that the ejection fraction (pumping capacity) improves by 30 to 40% as also the circulation system of the heart with the use of stem cell therapy, resulting in new blood vessels being created improving blood flow to the heart. This process generates new tissues in the heart muscle of the patient improving his energy levels to carry out the day-to-day tasks with more ease.

Our results are as follows:

| Condition | Parameter | %Change |
|---|---|---|
| Myocardial Infarction improvement | LeftVentricle Dysfunction | 25-45% |
| Primary Pulmonary Hypertension reduction | Systolic PAP(Pulmonary Artery Pressure) | 25-50% |
| Dilated Cardio Myopathy improvement | LV Ejection Fraction | 15-50% |

# STEM CELL THERAPY FOR SKIN DISORDERS

## Clinical Indications

### *Melanocyte Transplantation*

Stable vitiligo

Pigmented disorder

Amelanotic Nerves

Post-inflammatory hypo or depigmentation

### *Keratinocyte Transplantation*

Burn injury

Non-healing ulcers

Diabetic foots

## Role of Stemcure

We use a small piece of normal skin of the patient from the non-exposed area, either from the back of the ear lobe or buttocks (which is 1/10$^{th}$ of the diseased patch) and after that we develop the melanocyte in our special cell culture laboratory using different media and growth factors (which enhance the growth of the melanocyte cells). The prepared cells are then transplanted by expert clinician on to the affected area after removal of de-pigmented patch and the wound covered with collagen dressing. Usually, the dressing is removed after one week.

## Results

Usually treated area remains pink when the dressing is removed. Normally pigmentation is achieved within 3 to 4 months time with rare possibility of cosmetic side-effects. For full pigmentation, some patients may require more than one sitting to ensure a better texture of the skin.

# STEM CELL THERAPY FOR ORTHOPAEDIC DISORDERS

## Clinical Indications

### *Chondrocyte transplantation*

Osteochondritis Dissecans (OCD)

Cartilage Fracture

Sports Injury

### *Osteoblast transplantation*

Avascular necrosis

Osteoblastoma fibrocystic disease

Aneurismal bone cyst

Giant cell tumour (Gr-I)

Bone cysts.

## Role of Stemcure

### *Cartilage tissue culture*

After biopsy of the healthy knee cartilage, a very small portion is removed and cultured at our laboratory, developing millions of cells which are implanted into the defective area after removing the damaged tissue. Simultaneously, a small portion from a tissue covering the bone at the lower leg of the patient is removed and sutured over the defective area to ensure holding of the cells in place. These cells continue to multiply eventually integrating the surrounding cartilage to provide cells and tissues of the original cartilage.

## Results

Improvement in pain, locking, catching and limp following the implantation of cultured chondrocytes.

### *Bone Marrow Disorders*

**Signs and symptoms** Bone marrow disorder symptoms may be severe and life threatening or very subtle. Some chronic conditions may be discovered during yearly physical examinations, while other acute conditions cause symptoms related to too many or too few functioning blood cells. These symptoms include:

- Fever
- Weakness, fatigue, and pallor due to anaemia
- Weight loss

- Swollen lymph nodes, liver, spleen, kidneys, and testicles
- Bleeding and bruising
- Night sweats
- Bone and joint pain
- Frequent infections
- Headaches, vomiting, confusion and seizures (may occur when cells are increased in the brain or central nervous system).

**Treatments** There is usually no way to prevent bone marrow disorders. Some are a result of exposure to chemicals, due to previous radiation treatments, or due to rare genetic conditions but, for most cases, the cause is unknown. Treatment depends on the type of bone marrow disorder, its severity, and the symptoms it is causing. Treatment often changes over time. While acute leukaemia is sometimes curable, other bone marrow disorders are not. Leukaemias are frequently treated with chemotherapy or radiation. The goal of treatment is to drive the disease into remission and, if possible, to kill all of the abnormal blood cells, allowing normal cells to reproduce and restore blood cell function. Treatment for bone marrow disorders includes therapies for symptom relief. This may involve blood transfusions if anaemia is present or blood removal (therapeutic phlebotomy) if the bone marrow is making too many RBCs. Platelet transfusions may be necessary to control excess bleeding, while platelet removal (plateletpheresis to filter out platelets) may be required with excess platelet production. Frequent infections may necessitate antibiotic therapy and the administration of factors (granulocyte-colony stimulating factor or GM-granulocyte, monocyte stimulating factor, G-CSF) to stimulate the production of granulocytes and monocytes. Iron deficiencies may require iron supplementation. If the spleen becomes too swollen, it may need to be surgically removed. If the bone marrow disorder is severe and not responding to other therapies, a bone marrow transplant may be indicated. Bone marrow may be taken from the patient, "cleaned" of abnormal cells, and frozen to be reintroduced into the patient following treatment. It may also come from a suitable donor, most frequently a relative. New therapies for bone marrow disorders continue to emerge from research and clinical trials. Patients should talk to their doctors about the treatments that are best of their condition.

## Laboratory Tests

1. **Complete blood count (CBC) and differential** These are routine tests that are ordered to count the number and relative proportion of each of the different types of cells in the blood stream. They give your doctor information about the size, shape, and relative maturity of the blood cells present. CBCs and differentials are snapshots of what is happening in the body at the time the blood is taken. They are used to detect cell abnormalities, determine their importance, help diagnose their cause, and monitor their course and response to treatment. Irregularities in cell counts, such as elevated WBC counts or low RBC counts, may be due to bone marrow disorders, but they may also be due to a variety of other temporary or chronic conditions.

2. **Bone marrow aspiration/biopsy** If the doctor suspects a bone marrow disorder, then a bone marrow aspiration or biopsy is done to actually look at the cells and tissue in the marrow. When a laboratory technician examines the sample from your bone marrow under the microscope, the person can see the number, size, and shape of your red and white blood cells and platelet precursors (megakaryocytes), determine the proportions of mature and immature cells, see any overgrowth of fibrous tissue, and detect any cancer cells from cancers that may have spread to the marrow. Most bone marrow disorders can be diagnosed during this examination. Iron storage can also be estimated by a marrow examination, although if iron deficiency is suspected, other iron tests are usually ordered to confirm the condition and diagnose the cause.

3. **Flow cytometry** Flow cytometry is a technique that may be used to measure cell surface antigens and the DNA content of cells from the bone marrow, peripheral blood, or other body fluids. The technique involves incubating cells in liquid suspension after they have been tagged fluorescently. An instrument called a flow cytometer is then used to count and examine the cells for abnormalities.

4. **Genetic Tests** These tests look at several different types of genetic abnormalities in bone marrow cells or circulating white blood cells, including:

    - Chromosomal translocation; in which part of the chromosome is transferred to a different location, often to a different chromosome. It is associated with chronic myelogenous leukaemia, Burkitt's lymphoma, and many other leukaemias and lymphomas
    - Single gene changes; these are seen in some lymphomas and haemoglobin disorders.
    - Clonal populations of white blood cells; where many replicas of a particular WBC are produced. These provide diagnostic evidence for several myelodysplastic syndromes and lymphoid leukaemias.
    - Techniques involve removing the DNA from a specific cell population in the bone marrow or the blood, manipulating the DNA to find the correct gene, and then using special tests that look for one of the abnormalities listed above in the isolated gene.

5. **Spinal tap (lumbar puncture)** If leukaemia is found in the bone marrow, a spinal tap may be performed by the doctor as a follow-up test to look for leukaemic cells in the cerebral spinal fluid.

**Non-Laboratory Tests** X-rays are sometimes used to look for signs associated with some bone marrow diseases, such as masses of cells in areas like the chest, spleen, and liver.

# 14

# Mammary Stem Cell Assay

As such there is no known marker specifically expressed by mammary stem cells. However, there is a functional assay for rodent mammary stem cell activity, which is the transfer of whole populations or purified sub-populations, into cleared fat pads of juvenile rodents. Results obtained from fat pad assays suggest that stem cell activity does not change with age or parity of the female gland. Using flow cytometry, some investigators have purified cell populations enriched for their stem cell activity. In fact, they claim that just one cell can seed the growth of a completely new ductal tree, and this new tree includes a normal component of both mammary lineages.

## MAMMARY STEM CELLS AND BREAST CANCER

As for other tumours, it is very attractive to propose that somatic stem cells are a fast track precursor cell for mutation. If these cells have properties in common with other cancer stem cells, their growth, response to damage (for example, genotoxins) and death may be regulated by different means than the tumour majority and their successful elimination will need a separate treatment protocol.

## MARKERS OF STEMNESS

Stem cells are unspecialized cells that possess properties such as self-renewal, high potential for proliferation, and the ability to become a variety of cell types in the body. Self-renewal is the process of generating more stem cells. Differentiation is the process of becoming different cell types. Embryonic stem cells are isolated from very early stage embryos that possess the ability to develop into all different cell types in an individual. Adult stem cells, in contrast, are isolated from specific tissues that can only differentiate into cell types in that specific tissue under normal conditions. However, in rare occasions, the adult stem cells can become cell types of other tissues—a phenomenon known as transdifferentiation. Markers of stemness are special properties and molecular signatures. The special molecular signature can be a unique gene expression pattern or post-translational modifications that determine unique function of stem cells. The expression of these genes controls the establishment, survival, and maintenance of stem cells in undifferentiated states. Many growth factors and signalling molecules in the body regulate the specific gene expression pattern. The molecular signature of stem cells is commonly identified by the gene microarray method to globally assess the gene expression pattern of stem cells. By comparing the expression of genes in embryonic and adult stem cells, the common molecular signatures in adult and embryonic stem cells can be defined.

### Examples

OCT-4, SSEAs, CD133, ABCG2, and Nestin are several commonly used stemness markers. OCT-4 is one of the most commonly used markers for identifying embryonic stem cells. OCT-4 is required for embryonic stem cell self-renewal, to generate more stem cells, and for multiple lineage differentiation, to generate different cells in the body. Stage-specific embryonic antigens (SSEAs) are molecules that regulate cell-cell communication and interaction with intracellular structures. There are several SSEA proteins expressed in different tissue at various stages of development in embryonic stem cells and adult stem cells. CD133 is a sugar-attached protein that is expressed on the surface of blood stem cells and several tissue-specific stem cells such as neuron, breast, prostate, and colon stem cells. Recent studies suggest that CD133 is also expressed in cancer stem cells.

## Identification and Application

Stem cell markers are gene expression products that are uniquely related to stem cells. They are typically proteins or carbohydrates that are specifically expressed in stem cells. These markers are relatively specific and are usually expressed at a low level or are not expressed at all in terminal differentiated cells. Some of the stem cell markers are also expressed on many other differentiated cells in various tissues, and in tumour stem cells at different levels. So far, more than 100 stem cells markers have been identified in different tissues or at different stages of development, but the functional roles of most stem cell markers are not clear. Stem cell markers play a very important role in scientific research. Antibodies to stem cell markers are the most commonly used tools for the initial identification of stem cells in the tissue. Antibodies to stem cells are generated by immunizing animals with proteins associated with stem cell markers. Fluorescence-labelled antibodies to stem cell markers are used to identify stem cells or localize the stem cell niche. Several antibodies to stem cell markers can be used in combination to define different types of stem cells based on differential expression of these markers.

## Stem Cell Markers and Stem Cell Function

The criteria used to define the stem cell markers and the functional roles of the identified stem cell markers are heavily debated issues in stem cell research. The stem cell functional assay is a more reliable method for testing whether or not isolated cells are stem cells. Stem cell functional assays include *in vitro* cell proliferation and differentiation assays, as well as an *in vivo* assay to test whether or not isolated stem cells can generate the same types of tissues when transplanted to animals. The stem cell functional assay is time consuming and is difficult to perform in most cases. Thus, antibodies against stem cell markers become very important tools for the initial identification of stem cells.

## Identifying Stem Cells in a Mixed Population of Cells

Stem cells only make up a small proportion of the cells in any given tissue, and the expression of stem cell markers is tissue- and developmental stage-dependent. Analysis and interpretation of stem cell marker expression results is limited by purity and small numbers of stem cells can be isolated from a large population of differentiated cell types in the tissue. Stem cell markers are not exclusively expressed in stem cells; some differentiated cells also express stem cell markers at different levels. Most of the cells that express only one stem marker are usually not authentic stem cells. Multiple stem cell markers should be used in combination to define true stem cells in the tissue. Different tissues may express different types of stem cell markers. At present, there are no stem cell markers that can be used to identify all types of stem cells. In contrast, some stem cell markers can indicate stem cells in one tissue but have no predictive value in stem cells in other tissues. For example, stem cell antigen 1 is useful for the identification of mouse blood and prostate stem cells, but it is not consistently expressed in mouse breast ductal stem cells. To make things even more complicated, stem cell marker expression is highly dependent on the microenvironment for stem cell survival. The microenvironment for stem cells is called the stem cell niche. It is often difficult to compare stem cell marker expression without having the information on its microenvironment. Isolated stem cells will exit quiescent status and start to differentiate to different cell types under different microenvironments. Use of molecular genetics to manipulate these "stemness" genes help to clearly decipher the functional role of stem cell markers in the establishment and maintenance of stem cells.

## APPLICATION OF OTHER STEM CELL PROPERTIES

In addition to using an antibody to stain for stem cell markers, several other assays based on stem cell properties have also been used to identify stem cells. Stem cells are more quiescent and divide more slowly in comparison with differentiated cells in tissue. Therefore, stem cells can preserve labelled DNA for a longer period of time. Taking advantage of this property, DNA is labelled with molecular markers, and "slow cycling" cells are determined. The molecular mechanism of dye exclusion in stem cell is the specific expression of an ion channel that exports dye and other foreign reagents to cancer stem cells resistant to chemotherapeutic therapy. The dye exclusion property is not universally found in all stem cells; in some tissues, the dye exclusion cells are not stem cells. This method is valuable only when used in combination with other methods. DNA labelling retention, dye exclusion, and antibody staining can be used in combination to identify and isolate stem cells in the tissue. In addition, several signalling pathways, including TGF-beta, Wnt/beta-catenin, Bmi-1, Shh, and Notch, have been shown to be activated in stem cells or in the area where the stem cells are located. This property can also be used in stem cell identification and characterization. To accurately define stem cells in tissue is an important goal in stem cell research. The identification of stem cell markers unique to specific tissues and the determination of their relationship with functional stemness genes have significant effect on tissue stem cell research and potential clinical applications.

## USE OF STEM CELLS AS A THERAPY IN A FATAL SKIN DISEASE

Stem cells correct defect in a child's fatal skin disease. Researchers and clinicians have paved the way toward a cure for a young boy's genetic and fatal skin disease, recessive dystrophic epidermolysis bullosa (RDEB), by using a cord blood and bone marrow transplant. Children with RDEB have skin that is exquisitely delicate because of a genetic defect that results in the absence of a critical protein called collagen type VII, which anchors the skin and lining of the gastroinstestinal GI system to the body. Without collagen type VII, the skin of people with RDEB is extraordinary fragile; tearing and blistering occur with minimal friction. RDEB is associated with painful wounds, mutilating scaring, malnutrition due to the erosion of the oesophagus, and frequent infections, so to minimize the risk of tears and blistering those affected must have their entire body continuously wrapped in bandages. Up until now, there have been no successful treatments for children with RDEB and it is always fatal due to malnutrition, infection or eventually an aggressive skin cancer. In this trial, through the infusion of marrow stem cells obtained from a healthy donor, transplanted stem cells produce collagen type VII and correct the underlying genetic defect. The concept of using circulating stem cells to treat a skin disease like RDEB was introduced by Angela M.Christiano, Ph.D., professor of dermatology and genetics and development at Columbia University Medical Center, who has been involved in the study of the genetics of RDEB for more than 15 years.

Her laboratory performed the first genetics studies on the Liao family and determined the exact molecular basis for their individual case of RDEB. She has studied several gene therapy approaches for RDEB in the past before turning her attention to circulating stem cells. The groundbreaking research leading up to the clinical trial required that the investigators first showed evidence of a benefit of stem cell therapy in a mouse model of RDEB. Collaborating with Christiano, University of Minnesota researchers Jakub Tolar, M.D., Ph.D., and Bruce Blazar, M.D., discovered that certain stem cells found in bone marrow could correct the biochemical defect in RDEB. The mouse model of RDEB was generated specifically for the purpose of testing novel RDEB therapies, by the laboratory of Jouni Uitto, M.D., Ph.D., Professor and Chair, Department of Dermatology and Cutaneous Biology, at Thomas Jefferson University in Philadelphia. Analysis also was performed at Asahikawa Medical College in Asahikawa, Japan. Using the RDEB mouse, Drs.Tolar and Blazar showed that delivering marrow-derived stem cells in the bloodstream greatly lengthened the life expectancy of the mice, and also healed existing blisters. Using the positive results from the RDEB animal model, Dr.Wagner's team at UMN then designed a clinical trial to test stem cell therapy for RDEB for the first time in patients suffering from the disease. Drs.Christiano and Uitto are experts in the genetic basis of different forms of EB, and their work together has included identification of the human and mouse type VII collagen genes, establishing type VII collagen gene mutations as the molecular basis for dystrophic forms of EB, leading to the development of DNA-based diagnostic tests for EB and eventually to pre-clinical animal models, through research supported from Dystrophic Epidermolysis Bullosa Research Association of America (DebRa of America) and the National Institutes of Health. Dr.Christiano's ongoing work on stem cell therapy for RDEB and other disorders is supported by the New York State Foundation for Science, Technology and Innovation (NYSTAR). "The work announced today is the culmination of several years of preclinical work supported in part by the DebRA of America, specifically, the development of a mouse model that recapitulates the features of RDEB, including severe skin blistering at birth", said Mary Sprague, executive director of DebRA. "DebRa of America is pleased that its support to fundamental basic research in EB has facilitated this major achievement towards treatment, and eventually a cure, for RDEB".

# 15

# Stem Cells in Diabetes Mellitus

Diabetes mellitus (DM) is one of the prevailing hormonal diseases. It is often called a "non-infectious epidemic disease of the 21st century". Diabetes mellitus is characterized by a high blood sugar (glucose) level resulting from either insufficient insulin production in the body or body cell's improper response to the produced insulin. Insulin is a pancreatic hormone, the main function of which is regulating the blood sugar level. On the one hand, insulin helps body cells to absorb glucose and to turn it into energy. On the other hand, if glucose level is excessive, insulin causes liver and muscles cells to take up glucose from blood, transform it into glycogen and store it, thus enabling the body to receive glucose later. If insulin production or function disturbances occur, glucose concentration in the blood grows (hyperglycaemia), and the body cells lose their main source of energy. The only exception is insulin-independent cells, i.e., those cells that do not need insulin to supply glucose into a cell, for example, brain cells. DM falls into two main types: type 1 and type 2 diabetes mellitus. 85–90% cases of diabetes belong to type 2 DM. There are other diabetes types as well, but their course is similar to that of the above mentioned diabetes types.

## TYPE 1-DIABETES MELLITUS

Type 1 diabetes, also called pancreatic endocrine insufficiency, occurs when the body fails to produce insulin due to destruction of the insulin-producing beta cells in the pancreas. Pancreatic insufficiency starts with mass loss of insulin-producing cells leading to the drastic decrease of insulin. Type 1 diabetes mellitus is caused mostly by autoimmune attack against beta cells in the pancreas. Mass destruction of these cells can occur either without known cause or as a response to virus infections, toxins, pancreatic inflammatory diseases, traumas, tumours, or stress. Risk of developing type 1 diabetes also depends upon genetic predisposition. Due to the defects of some genes the body may start autoimmune aggression towards the cells of the pancreas; moreover, regenerative ability of beta cells decreases. Most affected are children, young people and middle-aged people (under 40). Type 1 diabetes mellitus symptoms develop with time. Traditional treatment consists of insulin injections. If diabetes is not treated, acute severe complications such as ketoacidosis and diabetic coma may develop that can lead to a patient's death.

## TYPE 2-DIABETES MELLITUS

In type 2 diabetes mellitus, also known as non-pancreatic insufficiency, insulin is produced in normal or even high amounts but response of the body cells to insulin is defective, which leads to insulin resistance. Sometimes, insulin resistance is accompanied by an insulin deficiency. Decrease of sensitivity of insulin-dependent tissues to insulin is either due to reduced number of insulin receptors on the membrane of insulin-dependent cells, or change in the quality of the receptors. Also, insulin structure can be damaged due to some genetic defects. Genetic predisposition increases the risk of developing type 2 diabetes two to six times. Excessive weight is one of the key factors of type 2 diabetes development. 80% of type 2 DM patients are stout. The risk of type 2 diabetes onset is two-fold higher in grade I obesity, five-fold higher in grade II obesity, and more than ten-fold higher in grade III obesity. Some other factors that can give rise to type 2 diabetes include advanced age, hypertension, sedentary life and bad habits. Unlike type 1 diabetes affecting mostly children, young and middle-aged people, type 2 diabetes develops mostly after 40. The disease progresses slowly. Chronic complications develop slowly, too. There is a very little risk of ketoacidosis development occurrence in type 2 diabetes mellitus, but being poorly compensated, type 2 DM can lead to diabetic coma.

## SYMPTOMS OF DIABETES

Clinical manifestations of diabetes type 1 and 2 are similar but their intensity is different. In type 1 diabetes mellitus, the symptoms are more distinct and may develop quite rapidly over a few weeks. On the opposite, type 2 diabetes is often diagnosed accidentally. Among the main symptoms of diabetes are frequent urination, persistent thirst due to significant water loss with urination, dryness in the mouth, increased hunger, weight loss (and even cachexy) in type 1 diabetes, or weight gain in type 2 diabetes. Secondary symptoms develop gradually and include itching of skin and mucous coats, general weakness, headaches and dizziness, inflammatory skin involvement, susceptibility to infectious diseases, eyesight impairment, decreased libido and sexual potency, etc. Intensity of symptoms varies in each case. It depends on the duration of the disease, extent of insulin secretion decrease as well as on the individual peculiarities of the sufferer.

## DIABETES-RELATED COMPLICATIONS

Stable blood sugar increase and impaired nutrition of body tissues cause a range of serious complications. Acute and chronic complications are described here. Among acute complications are diabetic ketoacidosis and diabetic coma. Chronic complications include diabetic macroangiopathy (damage of the large vessels), diabetic microangiopathy (damage of the small vessels such as arterioles, venules and capillaries), and diabetes neuropathy (nerve damage). Other complications include retinopathy that can result in blindness, renal damage (nephropathy) with gradual development of renal insufficiency, damage of the major arteries (atherosclerosis) of cardiac and cerebral arteries, as well as of the lower extremities arteries, soft tissues involvement such as trophic ulcers, and decreased immunity resulting in increased incidence of infectious-inflammatory processes.

## DIABETES TREATMENT WITH STEM CELLS

A fetal stem cell transplantation-based method has been developed (excluding pancreatic beta-cells) for diabetes mellitus treatment. This method has proven to be effective for diabetes type I and II treatment and is protected by patents of many countries. Stem cell treatment of diabetes results in pronounced hypoglycaemic effect, i.e., decrease of blood sugar level, allowing to reduce the dose of exogenous insulin by 50–70%. In more than half of cases diabetes treatment with fetal stem cells induces long-term clinical remission.

## Indications for Diabetes Treatment with Fetal Stem Cells

Stem cell treatment of diabetes is indicated at all stages of the disease. It is, however, the most effective in the cases of:

- New-onset insulin-dependent diabetes mellitus;
- Diabetes mellitus complicated by diabetic glomerulosclerosis, chronic renal failure (grade 1 and 2) and anaemic syndrome;
- Labile course of diabetes mellitus;
- Diabetes mellitus associated with complications due to infections and immune deficiency;
- Presence of treatment-resistant trophic ulcers of the soft tissues;
- Secondary sulfanilamide resistance and the need for patients with diabetes mellitus type II to transfer to insulin therapy.

## Effects of Diabetes Treatment with Fetal Stem Cells

Stem cell treatment of diabetes leads to significant improvement in patient's condition. In some cases at the early stages of the disease, it may result even in the full recovery. After the stem cell therapy, diabetes mellitus patients report normalization of immunological and haematological indices, reduced manifestations of micro- and macroangiopathy and trophic disturbances, restoration of workability. In case of treatment, the disease progression is hindered, and periods of remission become 2–3 times longer. Severity and frequency of diabetes complications decrease. Life quality and average life expectancy increase. Major effects of diabetes mellitus treatment with stem cells are listed as follows:

1. *Decrease of glycaemia in patients with new-onset insulin-dependent diabetes mellitus* Within 2–3 months after the stem cell treatment, in 100% of cases the dosage of administered insulin gradually decreased. With the average initial insulin dosage at 0.76_+0.06 units/kg/day, maximum dosage decrease amounted to 20-100% (41% on average), lasting for 14 to 90 days (59.0_+4.3 days on average). In 65% of cases, diabetes treatment resulted in clinical remission with exogenous insulin dosage dropping to less than 0.4 units/kg/day, or insulin therapy completely discontinued. Remission lasted 5 to 14 months.

2. *Improvement of psycho-physiological condition* The syndrome of psycho-physiological changes includes improvement of thinking and mental capacity, inburst of energy, disappearance of anxiety, depression decrease, etc. The syndrome has been noted in 48% cases.

3. *Immunity reconstruction* Diabetes treatment with fetal stem cells leads to restoration of immunological parameters. Lymphocyte count, T-lymphocytes and sub populations of T-lymphocytes normalize, B-lymphocytes count decreases by 30–60%. The immunocorrecting effect lasts for 3 to 8 months.

4. *Reconstitution of hematopoiesis* Within 1–1.5 months after the stem cell treatment of patients with diabetic glomerulosclerosis complicated by chronic renal failure (I, II and III degrees) and anaemic syndrome, restoration of hematopoiesis, normalization of erythrocyte count and haemoglobin were observed. This effect is maintained up to 12 months.

5. *Improvement of dystrophic disorders and trophic disturbances* After the transplantation of fetal stem cells, trophic ulcers disappear, microcirculation improves, and infectious skin lesions, fungus dermatitis, cutaneus lychenisation and lipoatrophic lesions decrease.

# 16

# Blood Disorders and Anemias

Blood disorders encompass a wide variety of diseases including anemias, cancer, hemophilias and immunodeficiencies. Blood disorders include various cancers such as leukemias, T cell lymphomas, B cell lymphomas, Burkitt's lymphoma, and Hodgkin's Disease. Treatment of cancer with gene and cell therapy is more fully discussed under cancer. Hemophilias, which are clotting disorders, and immunodeficiency diseases (e.g., Chronic Granulomatous Disease, Severe combined immunodeficiency, Wiskott-Aldrich Syndrome), Anemias include sickle cell anemia, beta and alpha thalassemias, and Fanconi anemia. Patients with anemia have low numbers of red blood cells or abnormal red blood cells. Inherited anemias are caused by a mutation in an essential gene involved in the development, function, or lifespan of the red blood cells or erythrocytes. Anemias susceptible to cell and gene therapy treatments includes sickle cell anemia, beta and alpha thalassemias, and Fanconi anemia. Anemias, like sickle cell anemia, beta and alpha thalassemias, can arise from mutations that affect the structure or the amount of one of the two globin chains of adults (alpha and beta). Sickle cell anemia arises from a mutation in codon 6 of beta globin and is called haemoglobin S. In the absence of oxygen, haemoglobin S clumps and causes the crescent shape of sickle red blood cells. Beta thalassemia can be caused by any one of over 200 mutations in the sequences that control production of the beta globin chains of haemoglobin. Similarly, alpha thalassemias have mutations that alter production of the alpha globin chain. The existence of alpha and beta thalassemias indicates that the level of both alpha and beta chains of haemoglobin need to be produced in the correct ratio and amounts to yield normal red blood cells. Fanconi anemia does not affect haemoglobin, but instead results from a mutation in one of 14 genes involved in the development of leukocytes, red blood cells and platelets. The proteins from these genes cluster together and help repair DNA after stress. Most patients with Fanconi anemia have a mutation in the *FANCA* gene (60–70%), the *FANCC* gene (9–15%) or the *FANCG* gene (8–9%). Current therapy that treats the immediate symptoms or crisis includes blood transfusions. Current cell therapy transfers bone marrow or hematopoietic stem cells from normal, human lymphocyte antigen (HLA) matched individuals to the patients. This cell therapy is the preferred treatment for these three types of anemias because it provides normal hematopoietic stem cells. The hematopoietic stem cells give rise to normal red blood cells. Gene and cell therapy is being developed for cases in which the patient does not have a HLA-matched sibling or donor. Basically, hematopoietic stem cells (HSC) are isolated from the patient. Gene therapy uses a vector to insert the correct gene into these hematopoietic stem cells from the patient. These corrected HSC are transferred into the patient so they can generate normal blood cells. Various methods that may increase the survival and function of the gene-corrected, transferred, hematopoietic stem cells in patients with anemia are being tested in the laboratory and animal models. The development of gene therapy for treatment of sickle cell anemia and beta thalassemia has encountered several challenges:

- First individuals can vary in the severity of their disease. In the case of thalassemias, although individual patients may have one of many different mutations, generally the deficiency in globin production is severe and any significant level of added expression is likely to be therapeutic. Patients with Fanconi anemia may have the mutation in one of 14 different genes. Such patients, although diagnosed with the same disease, but having different mutations often require a gene therapy reagent specific for the gene in which their mutation is located.

- Second, the development of vectors that express an adequate amount of beta globin has been challenging. For example, early vectors produced beta globin for a short time only and stopped its production or were silenced. To overcome premature silencing of globin expression, scientists have added insulators or sequences which block the effects of neighboring genes. Scientific challenges include the identification of the elements that permit silencing and those regions that are sensitive to silencing.

- Third, scientists are developing new vectors such as Lentiviral vectors and adeno-associated virus type 8 (AAV-8) to improve long term expression of the introduced gene and to achieve targeting of the gene therapy to the right cell types. Recent clinical trials are testing the ability of new vectors to correct the defect in isolated human hematopoietic stem cells from anemia patients. These modified human cells are being transferred into special mice and monitored for their ability to make normal red blood cells. Other clinical trials are testing various approaches that can increase the survival of the gene-corrected cells transferred to the patients.

- A fourth challenge is harvesting a sufficient number of stem cells from the patients. Several reagents increase stem cells numbers in healthy individuals. One recent trial is testing reagents that may increase the number of stem cells in the blood of anemia patients.

# 17

# Immunodeficiency Diseases

Immunodeficiency is a defect in the immune system that prevents immune resistance to infectious disease and some forms of cancer. There are two types of immunodeficiency: the primary or congenital form, caused by genetic defect in one of several essential genes involved in maturation of the immune response; and the acquired or secondary immunodeficiency caused by chemotherapy, some forms of cancer or chronic infections such as AIDS. Congenital diseases are caused by a mutation in any one of several genes essential to the development of a fully active immune response. Although congenital immunodeficiency was initially treated by transplantation of allogeneic hematopoietic stem cells, the past 20 years has shown that these conditions are correctable by gene therapy. One noted patient with severe combined immune deficiency (SCID) lived in an enclosed room for several years to protect him against infection and therefore this disease has been nicknamed "Bubble Baby Disease". The current treatment of choice for immunodeficiency diseases is the transplantation of bone marrow or hematopoietic stem cells from matched siblings. However, many patients do not have a matched sibling who can serve as a donor. One type of SCID is caused by the lack of a protein (adenosine deaminase or ADA-SCID) which has been treated by periodic administration of missing ADA enzyme that results in clinical improvement, but requires an expensive lifelong treatment. Gene-cell therapy strategies that insert a normal copy of the defective gene for ADA into the patient's own hematopoietic stem cells are being tested as effective treatment in different forms of SCID. The current status of gene and cell therapy for the immunodeficiency diseases treated to date is summarized in the following section.

## ADA-SCID

A mutation or deletion in the adenosine deaminase (ADA) gene can cause adenosine deaminase-severe combined immunodeficiency (ADA-SCID). ADA-SCID patients fail to make T cells, B cells and NK (Natural Killer) cells, experience recurrent infections and fail to thrive. Although the first trials detected gene corrected T cells in the patients, the number of "corrected" and functional T cell levels were low. The experimental approach was improved by pre-treating the patients with a low dosage of busulfan, a chemotherapy drug which facilitated the engraftment of the ADA-corrected hematopoietic stem cells in the bone marrow. The transplanted gene modified stem cells, provided gene corrected lymphocytes and most patients treated in recent years have remained well and have not needed ADA replacement enzyme therapy. Improved vectors, based on lentiviral vectors, are currently in development.

## X-LINKED SCID

Deletions or mutations of the common gamma chain (gc) found in the cell surface receptors for several cytokines (hormones that activate the immune cells) causes X-linked severe combined immunodeficiency (X-linked SCID, also called SCID-X1). The immune cells of patients with X-linked SCID cannot respond to the several essential cytokines (IL-2, IL-4, IL-7, IL-9, IL-15, IL-21) needed for these cells to survive and fight infections. Thus, patients with X-linked SCID lack T cells and NK cells and the functions of their B cells. The basic gene-cell therapy strategy for treatment of X-linked SCID was similar to the approach shown for the ADA-SCID form of disease, except that they have not required treatment with chemotherapy. A retroviral vector carried the normal gamma chain receptor gene into some of the hematopoietic stem cells from the bone marrow of X linked-SCID patients. The cells were transfused back into the patients. These patients enrolled in two trials experienced higher, functional T cell and

NK numbers. New T cells are still maturing in these patients ten years after the initial gene therapy. The results suggest that therapy is more efficacious in younger patients. Patients treated with the first generation retroviral vector based gene therapy developed a leukaemia that was related to the treatment. They were treated for this complication and survived with continued corrections of their immune system. Few patients succumbed to the leukaemia. However, these unfortunate events have spurred research on understanding the mechanisms of viral induced leukaemia, the development of safe vectors, and the improvement of safety standards. New clinical trials are enrolling patients using these advances.

## CHRONIC GRANULOMATOUS DISEASE (CGD)

Chronic granulomatous disease is caused by the inability of white blood cells to kill invading fungi and bacteria, because they lack a normal gene needed for the production of the germ-killing superoxide and toxic oxygen metabolites. A defect in the *gp91phox* gene occurs in approximately 70% of CGD patients. The *gp91phox* gene is located on the X chromosome and so this form of CGD affects males. In clinical trials, hematopoietic stem cells (HSC) from CGD patients were treated with gene therapy to insert the normal *gp91phox* and then were transferred back into the CGD patients. As for the ADA-SCID patients, low dose busulfan, which facilitates the engraftment of gene-modified HSC, was given to improve the efficacy. Remarkably, these patients were able to fight off serious infections that had been resistant to all forms of medical management, including new generations of antibiotics. Some patients developed abnormalities in their bone marrow resulting in the under production of a class of blood cells, a condition called myelodysplasia. New gene transfer vectors have been developed that maintain long term expression of the *gp91phox* gene and have an improved safety profile in this type of cells.

## WISKOTT-ALDRICH SYNDROME

The gene mutated in Wiskott-Aldrich syndrome codes for WASP, a key protein that regulates the shape and function of several types of blood cells. The defects in WASP affect platelet numbers and size, the activity of several types of white blood cells and antibody production. Patients with severe Wiskott-Aldrich syndrome have low platelet counts (which may lead to severe bleeding), eczema, and immunodeficiency. They are highly susceptible to infections and autoimmune disorders. Clinical trials of gene therapy for Wiskott-Aldrich syndrome have been initiated recently, and this therapy appears to be effective in correcting the function of immune cells. In summary, gene-cell therapy protocols have shown promise in treatment of patients with immunodeficiency diseases. Challenges imposed by complications that occurred in trials in CGD and X-linked SCID have been met with the development of improved vectors that have been shown to be safer in pre-clinical studies. Ongoing clinical trials hope to provide effective and safe treatments for more children.

# 18

# An Overview on Stem Cells

Stem cells are the master cells of the human body. These cells differ from other cells by their ability to self-generate themselves and the ability to differentiate into other cell types.

Stem cells are at the centre of a new field of science called regenerative medicine. All stem cells regardless of their source have three general properties:

(1) They are capable of dividing and renewing themselves for long periods;

(2) They are unspecialized; and

(3) They can give rise to specialized cell types.

Self-regeneration is the ability of stem cells to divide and produce more stem cells. During early development, the cell division is symmetrical, i.e., each cell divides and gives rise to daughter cells each with the same potential. Later in development, the cell divides asymmetrically with one of the daughter cells and also produced stem cells and other more differentiated cells (such as brain cell, blood cell, etc.)

When stem cells are being transplanted into the body and arrive (through migration or through direct injection) into the injured part of the brain which is being targeted for tissue regeneration. These cells come in contact with growth chemicals in the body and these chemicals program the stem cells to grow into the tissue surrounding it. Stem cells are typically broken into four types.

- Embryonic stem cells: Stem cells taken from human embryos
- Fetal stem cells: Stem cells taken from aborted fetal tissue
- Umbilical stem cells: Stem cells taken from umbilical cords
- Adult stem cells: Stem cells taken from adult tissue

Stem cells are defined by two properties. Firstly, they can 'self-renew', that is they can divide and give rise to more stem cells of the same kind. Secondly, they can mature or 'differentiate' into specialized cells that carry out a specific function, such as in the skin, muscle, or blood.

There are many different types of stem cells. These include embryonic stem cells that exist only at earliest stages of development, and various types of 'tissue-specific' stem cells (sometimes referred to as 'adult' or 'somatic' stem cells) that are found in various tissues in our bodies. Recently, cells with properties similar to embryonic stem cells, referred to as induced pluripotent stem cells (iPS cells), have been engineered from specialized cells such as skin cells.

## What is Stem Cell Therapy?

A stem cell therapy is a treatment that uses stem cells, or cells that come from stem cells, to replace or to repair a patient's cells or tissues that are damaged. The stem cells are either pooled into the blood, or transplanted into the damaged tissue directly or even recruited from the patient's own tissues for self repair.

## What are Embryonic Stem Cells?

Embryonic stem cells are pluripotent cells, in other words, cells that can become any type of adult cell. Embryonic stem cells are obtained from a blastocyst, which in humans, usually forms five days after the fertilization of an egg. The blastocyst is a mainly hollow ball, approximately consisting of 150–200 cells and at this stage, there are no organs or blood. However, inside of the blastocyst is the inner cell mass from which embryonic stem cells are grown. Human embryonic stem cells are derived primarily from blastocysts that were no longer needed in the in vitro fertilization ( IVF) process.

## What are Adult Stem Cells?

Adult stem cells are any cells that are found in an adult organism and contains two properties; firstly the ability to divide and create another cell like itself and, secondly the ability to create another cell more differential from itself. Adult stem cells are also called tissue specific or somatic stem cells and they do not have the same potency or ability to change as embryonic stem cells do

## What are hRPE Stem Cells?

hRPE ( Human Retinal Pigment Epithelial) Stem Cells are adult multipotent stem cells that are found in the retina of the human eye. The original function of these special cells is producing dopamine that participates in the pigmentation process inside the retina. Due to their function, hRPE cells are a new and most promising solution for Parkinson's patients who suffer from degeneration of their dopamine-producing cells. Dopamine is a neuro-transmitter in our brain that supports our movement fluidity and continuity. hRPE cells can replace degenerated dopamine-producing neurons in the nigrostriatal pathway and provide a source of increased dopamine production to promote functional recovery in Parkinson's patients.

The hRPE cells are derived from a single donor, and then cultured in the laboratory to produce the amount and the type of cells fitting for implantation. The cells are then being injected within a single operation into 4 targets in the putamen area in the brain, using MRI 3D image guided stereotactic technique.

While the treatment cannot completely remove all symptoms of the disease, it can greatly reduce the tremors, decrease muscle tension and rigidity, improve movement continuity, movement initiation and balance, increase muscle strength, improve slurred speech, reduce the freeze ups (stopped functions of the body) and strengthen the overall body.

Every patient who wishes to apply for this treatment should undergo a thorough evaluation process.

## VISTA STEM CELL: CHINA—PICTORIAL MODEL OF A STEM CELL CENTRE

Stem cell therapy has come a long way over the last ten years, despite concerns expressed by some western governments over approaches to research. One place which has not suffered from these setbacks is China. Professionals at the Vista Stem Cell facilities in Beijing have been administering stem cell treatments using embryonic stem cells for nearly ten years with ever improving results. Their sophisticated techniques and experience ensure patients receive the highest quality stem cell therapy and an alternative to existing treatments.

Vista Stem Cell staff use the strictest protocols to ensure that all their stem cells are disease-free and healthy. Their mission is to improve the quality of life of all their patients and enable to gain control over their life.

### Aims, Values and Promises at Vista Stem Cell

- They aim to provide the best stem cell treatment along with an excellent follow-up service.
- They adhere to GMP standards to ensure that the stem cells given to patients are safe and effective. They also believe in being honest and giving patients realistic expectations as to how successful their course of treatment will be.
- They promise to provide us with the highest standard of care and support. They do not promise to cure the disease, but they do promise to do their best to ensure better quality of life after the treatment.

### Stem Cell Treatments Available from Vista Stem Cell

Due to the ongoing nature of research, the number of diseases that Vista Stem Cell can treat with embryonic stem cells is constantly changing. Given in the following section is a current list of diseases that they can treat and have treated successfully.

| **Neurological diseases** | **Cancer** |
|---|---|
| Myelitis | Liver |
| Stroke | Stomach |
| Epilepsy | Lung |
| Autism | Breast |
| Cerebral Palsy | Pancreatic |
| Cerebral Haemorrhage | Ovarian |
| Cerebral Infarction | Uterine |
| Spinal cord Injury | Oesophageal |
| Encephalopathy | Colorectal |
| Motor Neuron Disease | Rectal |
| Alzheimer's Disease | Tongue |
| Encephalitis | Oral and Bile duct cancer |
| Cerebral Atrophy | Lymphoma |
| Multiple Sclerosis | Multiple Sclerosis |
| Parkinson's Disease | Brain tumours and other tumours |
| Asynergia | |
| Myasthenia Gravis | Leukaemia |
| Muscular Dystrophy | |
| Polyneuropathy | **Immunesystem diseases** |
| Traumatic Brain Injury | Sjogren |
| Torsion Spasm | Lupus |
| Optic Atrophy | Psoriasis |
| Peripheral Neuropathy | Thromboangitis obliterans |
| Optic Nerve Hypoplasia | **Cardiovascular diseases** |
| Ataxia | Myocardial Infarction |
| ALS | |
| Glaucoma | |
| | **Others** |
| **Endocrine Diseases** | Femoral Head Necrosis |
| Diabetes and complications | Kidney disease |
| | Liver disease |

## The Treatment Clinic

The Navy General Hospital is a multi-purpose hospital that provides a full range of medical care and serves an important role as a university teaching and research hospital. Vista Stem Cell's stem cell treatment facility is based in the Department of Neurosurgery and is headed by a team of experience doctors.

The team is a world leader in stereotactic brain surgery and has pioneered a number of ground-breaking methods in this area of medicine.

# FREQUENTLY ASKED QUESTION (FAQ'S)

## What are stem cell lines and why do researchers want to use them?

A stem cell line is a group of cells that all descend from a single original stem cell. Cells in a stem line keep growing but do not differentiate into specialized cells. Ideally, they remain free of genetic defects and continue to create more stem cells. Clusters of cells can be taken from a stem cell line and frozen for storage or shared with other researchers. That way, researchers do not have to get stem cells from an embryo itself.

## What is stem cell therapy (regenerative medicine) and how does it work?

Stem cell therapy, also known as regenerative medicine, is the replacement of diseased, dysfunctional or injured cells with stem cells or their derivatives. It is somewhat similar to the organ transplant process but uses cells instead of organs.

Researchers grow stem cells in the laboratory. These stem cells are manipulated to make them specialize into specific types of cells, such as heart muscle cells, blood cells or nerve cells. This manipulation may involve changing the material in which the stem cells are grown or even injecting genes into the cells. The specialized cells could then be implanted into a person. For example, if the person had heart disease, the cells could be injected into the heart muscle. The healthy, transplanted heart cells could then contribute to repairing defective heart muscle. In fact, researchers have already shown that adult bone marrow cells guided to become heart-like cells which can repair heart tissue in mice, and much more research is ongoing.

## Have stem cells already been used to treat diseases?

Yes, stem cell transplants, also known as bone marrow transplants, have been performed in the United States since the late 1960s. These transplants use adult stem cells.

Adult stem cells are being tested in other applications, including a number of degenerative diseases, such as heart failure. Stem cells from umbilical cord blood have been successfully used in clinical trials to treat cancer and blood-related diseases.

The Food and Drug Administration has approved an initial safety study using an embryonic stem cell therapy for the treatment of serious spinal cord injuries.

## What are the potential problems with using embryonic stem cells in humans?

Embryonic stem cells could grow irregularly—something that has been reported in animal experiments—or travel to a part of the body where they are not intended to go. They also might trigger an immune response in which the recipient's body attacks the stem cells as foreign invaders, or simply fail to function normally, with unknown consequences. Researchers are working on ways to avoid these possible complications.

## What is therapeutic cloning and what benefits do they offer?

Therapeutic cloning is a technique to create versatile stem cells independent of fertilized eggs. In these techniques, the nucleus, which contains the genetic material, is removed from an unfertilized egg. The nucleus is also removed from a somatic cell of a donor. This donor nucleus is then injected into the egg, replacing the nucleus that was removed, a process called nuclear transfer. The egg is allowed to divide and soon forms a blastocyst. This creates a line of stem cells that is genetically identical to the donor's cells. This technique is also called somatic cell nuclear transfer.

Some researchers believe that stem cells derived from therapeutic cloning may offer benefits over those from fertilized eggs because they are less likely to be rejected once transplanted back into the donor.

### Has therapeutic cloning in people been successful?

No. Researchers have not been able to successfully perform therapeutic cloning with humans, despite success in a number of other species.

### What does the future hold for stem cell therapy?

Researchers say the field has promise. Stem cell transplants using adult stem cells continue to be refined and improved and researchers are discovering that adult stem cells may be somewhat more versatile than originally thought, which means they may be able to treat a wider variety of diseases. Studies using embryonic stem cells or embryonic-like stem cells to regenerate tissue and organs in people are just getting started. Researchers are enthusiastic about the potential for regenerative medicine treatments.

## TYPES OF STEM CELL IN USE

What kind of stem cells are usually used? Is it umbilical cord, embryonic or adult stem cells? Where are they obtained from?

The type of stem cells we choose depends on the disease that the patient suffers from, and the medical plan. For Parkinson's diseases, we use human retinal pigment epithelial (hRPE) cells from adults. For cerebral palsy and strokes, we use self-stem cells. All stem cells come from the Peking University Stem Cell Research Center, which is the largest and most qualified stem cell laboratory in China, where the cells had undergone rigorous tests according to our highly strict regulations.

# CORD BLOOD CELLS AND RELATED CELLS

Umbilical cord is a special bond between the mother and the child that nourishes, protects and plays a vital role in the growth and development of the child during pregnancy. Now we can preserve this bond for life time for future health of the loved ones. Removing of the umbilical cord is not harmful to the baby and the blood would normally be thrown away as medical waste.

## Quick Updates

Cord blood is obtained from the umbilical cord at the time of childbirth, after the cord has been detached from the newborn and is extracted from the fetal end of the cord. Collected cord blood is cryopreserved (at –196°C) and then stored in a cord blood bank for future transplantation. Stem cell banking offers a unique opportunity for storage of umbilical cord blood stem cell and umbilical cord mesenchymal stem cell. Stem cell are unspecialized cells that can self renew indefinitely through cell division and also differentiate into more mature cells with specialized function. Cord blood stem cells are currently used in treatment of more than 75 life threatening diseases.

## Quick Links

Baby cell brings absolutely free ANC program first time in India. Baby cell is listed on www.parentsguidecordblood.org

To know more about this, View Online Brochure

Baby cell Preferred Articles

Stem cell banking is an easy, painless and non-invasive procedure. And ever a need arises in the future, these stem cells can be used as a powerful therapy. It takes just 10 minutes to bio-insure the lives of the loved ones. Baby cell, operated by Regenerative Medical Services Pvt. Ltd., is India's first Umbilical Cord Blood Bank to store samples at internationally accredited laboratory. We are a dedicated team of doctors and scientists pioneering new innovations in Regenerative Medicines. This scientific expertise has enabled us to assure utmost safety to the samples preserved.

## Umbilical Cord Blood and Cord Tissue

Umbilical cord blood, commonly called 'cord blood' is the blood that remains in the umbilical cord and placenta following birth of your baby. The umbilical cord is routinely discarded with the placenta. The baby's umbilical cord blood is loaded with stem cells, which are genetically unique to the baby and family.

The cord tissue is the tissue constituting the umbilical cord after removal of the cord blood.

Stem cells are the "master" cells of the body. They are undifferentiated, "blank" cells that do not yet have a specific function. Characteristically, stem cells have a high capacity for self-renewal and can regenerate and turn into the cells that form all types of tissues, organs and systems in the human body. They have the ability to replace or repair damaged cells throughout the body.

## Cord Blood

The stem cells found in cord blood are "Hematopoietic stem cells" which are the building blocks of the baby's blood and immune system and most readily replicate into the following:

1. Red Blood Cells which carry oxygen to all the cells in the body,
2. White blood cells which fight infection and,
3. Platelets which aid in clotting in the event of injury. The ability of cord blood stem cells to differentiate or change into other types of cells in the body is a new discovery that holds significant promise for improving the treatment of some of the most critical diseases like leukaemia, thalassemia, juvenile diabetes, cerebral palsy, etc. Cord blood stem cells have the potency to treat over 75 serious ailments. With advancing research and clinical trials worldwide, every day holds a promise for a new disease to be treated using cord blood stem cells.

## Cord Tissue

The baby's cord tissue is a valuable source of stem cells that could differentiate into various types of tissues like heart cell, muscle cell, etc. and be used to treat many lifestyle and tissue degenerative diseases and medical conditions. The umbilical cord tissue stem cells are "mesenchymal stem cells" which have been found to have the potency to treat tissue / organ related ailments like diabetes, heart ailments, osteoporosis, etc.

List of diseases that can be potentially treated using cord tissue stem cells

1. Chronic Obstructive Pulmonary Disease
2. Cardiac Arrhythmia
3. Crohn's Disease
4. Graft vs Host Disease
5. Critical Limb Ischemia
6. Multiple Sclerosis
7. Diabetes

It is important to preserve both the cord blood and cord tissue that yields stem cells so as to ensure that the child gets a complete recovery from most of the ailments what the future may promise. There are three major sources where stem cells are commonly found, they are:

1. Bone Marrow
2. Umbilical Cord Blood and Cord tissue
3. Menstrual blood

Stem cells are widely used through a procedure called 'stem cell transplantation' to regenerate a patient's blood and immune system and tissues.

# TYPES OF CELLS

The body is made up of about 200 different kinds of specialized cells such as muscle cells, nerve cells, fat cells and skin cells. All specialized cells originate from stem cells. A stem cell is a cell that is not yet specialized. The process of specialization is called differentiation and once the differentiation pathway of a stem cell has been decided, it can no longer become another type of cell. A stem cell that can become every type of cell in the body is called pluripotent whilst a stem cell that can become only some types of cells is called multipotent. Stem cells are found in the early embryo, the foetus, placenta, umbilical cord, and in many different tissues of the adult body.

Stem cells are often divided into two groups: tissue specific stem cells (often referred to as adult stem cells) and pluripotent stem cells (including embryonic stem cells and induced pluripotent stem cells). Tissue specific stem cells are derived from, or resident in, fetel or adult tissue, and can usually only give rise to the cells of that tissue, thus they are considered multipotent. Embryonic stem cells, derived from a small group of cells within the very early embryo, and their new counterpart induced pluripotent stem (iPS) cells are considered pluripotent as they can become every type of cell in the body.

## Tissue Specific Stem Cells

Tissue specific stem cells are undifferentiated cells found in the tissues and organs of the body. They are capable of self-renewal. Their differentiation is mainly restricted to forming the cell types of that tissue or organ. The chief role of tissue specific stem cells is to maintain and repair the tissue in which they are found. Skin stem cells, for example, give rise to new skin cells, ensuring that old or damaged skin cells are replenished.

It now appears that all tissues probably contain adult stem cells. Most tissues contain only tiny numbers of stem cells. The exception is bone marrow and umbilical cord blood which contains relatively high numbers of stem cells. In each tissue, adult stem cells are used to produce new mature cells as old ones die in the natural processes of ageing. They may also be activated by disease or injury.

Due to their small numbers, isolation of adult stem cells is difficult, but they have been successfully isolated from the brain, bone marrow, blood, muscle, skin, lung, pancreas and liver. The majority of research has been carried out on hematopoietic stem cells isolated from bone marrow and umbilical cord blood and on mesenchymal stem cells which can also be sourced from the bone marrow and some other tissues. Mesenchymal stem cells are the stem cells that form our fat, muscle, bone and cartilage and they can also differentiate into nerve cells.

Hematopoietic stem cells are the stem cells from which all blood cells and many of the cells of our adult immune system are derived. These are the stem cells with the longest history of clinical use in treating disorders such as leukaemia via bone marrow transplants. There has recently been much interest in whether hematopoietic stem cells can be made to differentiate into non-blood cells, such as heart muscle cells or even nerve cells.

Mesenchymal stem cells can be found in the bone marrow but are also found in several other sites in the body such as the placenta. Mesenchymal stem cells are particularly interesting to researchers because in addition to their capacity to differentiate into the multiple cell types listed above, they also have anti-inflammatory and immune-suppressing properties. This means that mesenchymal stem cells could be used as a therapy for diseases caused by immune attack on specific tissues.

Umbilical cord blood cells are a type of tissue-specific stem cell. Blood can be collected from the umbilical cord of a newborn baby shortly after birth. This blood is rich in hematopoietic stem cells that can be used to generate blood cells and cells of the immune system. Cord blood stem cells may be used to treat a range of blood disorders and immune system conditions such as leukaemia, anemia and autoimmune diseases. Once collected, cord blood can be stored in a cord blood bank for future use as a potential source of stem cells for transplant.

***Mesenchymal stem cells*** Mesenchymal stem cells (MSCs) have been used in cell-based therapy in various disease conditions such as graft-versus-host and heart diseases, osteogenesis imperfecta, and spinal cord injuries, and the results have been encouraging. However, as MSC therapy gains popularity among practitioners and researchers, there have been reports on the adverse effects of MSCs especially in the context of tumour modulation and malignant transformation.

The mechanism of the modulatory behaviour and the tumorigenic potential of MSCs, warrant urgent exploration, and the use of MSCs in patients with cancer awaits further evaluation. However, if MSCs truly play a role in tumour modulation, they can also be potential targets of cancer treatment.

**Umbilical Cord-derived Mesenchymal Cells** There are growing interests in the use of stem cells, e.g., embryonic stem cells, mesenchymal stem cells, hematopoietic stem cells, etc., in treating human congenital diseases caused by genetic mutation and acquired disease. Especially, transplantation of mesenchymal stem cells has drawn much attention in regenerative medicine. Since mesenchymal stem cells (MSCs) are first isolated from bone marrow, these plastic adherent cells have been identified and cultivated from many connective tissues, e.g., umbilical cord, amniotic membrane, cartilage, adipose tissue, and conjunctiva. MSCs are multipotent and are capable of differentiating into various progenitor cells that form connective tissues such as bone, bone marrow, cartilage, adipose tissue, etc. Thus, there have been growing interests in the application of MSC transplantation in regenerative medicine to repair and restore normal function of diseased and injured tissues. For example, bone marrow MSCs have been used in treating diseases in animal models as well as in human clinical trials in areas such as autoimmune diseases, solid organ allograft survival, hepatic cirrhosis kidney diseases, neuro- and muscle- degenerative diseases, myocardial infarction, and spinal cord injury. Umbilical cord-derived mesenchymal stem cells (UMSCs ) from newborns are presumably young and therefore are likely to have a more potent proliferation and differentiation capability. Other advantages to the use of UMSCs is that the availability of umbilical cords from newborn babies is almost unlimited and the cells can be expanded and stored in liquid nitrogen in a tissue bank and thawed for use. However, the utilization of human UMSCs in treating disease is surprisingly rare in comparison to the use of bone marrow MSCs. Only a limited number of publications have reported on the success of using human UMSCs to treat liver fibrosis and spinal cord injury in animal models.

**Characteristics of UMSCs isolated from human umbilical cord** Two populations of stem cells, UMSC and umbilical cord-derived hematopoietic stem cells (UHSCs) were isolated from human umbilical cord and cord blood, respectively. To characterize the stem cell, flow cytometry was used to determine the expression pattern of CD markers. The results showed that more than 95 percent of the fibroblast-like plastic-adherent cell expressed characteristic CD markers of mesenchymal stem cells, e.g., D105/SH2 (endoglin) CD73/SH3(NT5E), CD90(Thy1), CD29 (integrin B1) CD44 (HA receptor) CD13 (alanine aminopeptidase); and less than 1 percent of cells were positive for hematopoietic cell CD markers, e.g., CD45 (leukocytes), CD#4 (hematopoietic stem cells), CD14 (co-receptor of TLR4 and MD-2) HLA-DR(MHC-II) , and CD31 (endothelial cells)

The plastic adherent UMSCs were multipotent and capable of differentiation into adipocytes, osteoblasts and chondrocytes when they were cultured in adipogenic, osteogenic, and chondrogenic medium, respectively. To demonstrate differentiation into these cell types, staining with oil-red O, alkaline phosphates, and toluidine blue were performed. Oil-red O staining showed large amounts of lipid droplets indicating induction of adipogenesis. Osteogenesis was confirmed by expression of alkaline phosphatase while the synthesis of chondroitin sulfate verified by toluidine blue staining demonstrated induction of chondrogenesis.

To assess the percentage of CD34 positive cells after the volume reduction of cord blood, flow cytometry was employed to examine the characteristics of cord blood. The results showed that the cord blood cells were positive to CD34, CD45, CD59, CD133, and BCRP1.

To verify the expression pattern of UMSCs and UHSCs prior to transplantation, immune fluorescent staining was performed with selected CD makers for MSCs and HSCs, as well as to keratocan and lumican. The results showed that both human UMSCs and UHSCs did not express keratocan and lumican, and UMSCs were CD34 negative.

## IDENTIFICATION OF STEM CELLS

Although there is not complete agreement among scientists of how to identify stem cells, most tests are based on making sure that stem cells are undifferentiated and capable of self-renewal. Tests are often conducted in the laboratory to check for these properties.

One way to identify stem cells in a laboratory, and the standard procedure for testing bone marrow or hematopoietic stem cell (HSC), is by transplanting one cell to save an individual without HSCs. If the stem cell produces new blood and immune cells, it demonstrates its potency.

Clonogenic assays (a laboratory procedure) can also be employed *in vitro* to test whether single cells can differentiate and self-renew. Researchers may also inspect cells under a microscope to see if they are healthy and undifferentiated or they may examine chromosomes.

To test whether human embryonic stem cells are pluripotent, scientists allow the cells to differentiate spontaneously in cell culture, manipulate the cells so they will differentiate to form specific cell types, or inject the cells into an immune suppressed mouse to test for the formation of a teratoma (a benign tumour containing a mixture of differentiated cells).

## ONGOING RESEARCH WITH STEM CELLS

Scientists and researchers are interested in stem cells for several reasons. Although stem cells do not serve any one function, many have the capacity to serve any function after they are instructed to specialize. Every cell in the body, for example, is derived from first few stem cells formed in the early stages of embryological development. Therefore, stem cells extracted from embryos can be induced to become any desired cell type. This property makes stem cells powerful enough to regenerate damaged tissue under the right conditions.

## APPLICATION OF STEM CELLS

### Organ and Tissue Regeneration

Tissue regeneration is probably the most important possible application of stem cell research. Currently, organs must be donated and transplanted, but the demand for organs far exceeds supply. Stem cells could potentially be used to grow a particular type of tissue or organ if directed to differentiate in a certain way. Stem cells that lie just beneath the skin, for example, have been used to engineer new skin tissue that can be grafted on to burn victims.

## Brain Disease Treatment

Additionally, replacement cells and tissues may be used to treat brain disease such as Parkinson's and Alzheimer's by replenishing damaged tissue, bringing back the specialized brain cells that keep unneeded muscles from moving. Embryonic stem cells have recently been directed to differentiate into these types of cells, and so treatments are promising.

## Cell Deficiency Therapy

Healthy heart cells developed in a laboratory may one day be transplanted into patients with heart disease, replacing the heart with healthy tissue. Similarly, people with type I diabetes may receive pancreatic cells to replace the insulin-producing cells that have been lost or destroyed by the patient's own immune system. The only current therapy is a pancreatic transplant, and it is unlikely to occur due to a small supply of pancreas available for transplant.

## Blood Disease Treatments

Adult hematopoietic stem cells found in blood and bone marrow have been used for years to treat diseases such as leukaemia, sickle cell anemia, and other immune deficiencies. These cells are capable of producing all blood cell types, such as red blood cells that carry oxygen to white blood cells that fight disease. Difficulties arise in the extraction of these cells through the use of invasive bone marrow transplants. However hematopoietic stem cells have also been found in the umbilical cord and placenta. This has led some scientists to call for an umbilical cord blood bank to make these powerful cells more easily obtainable and to decrease the chances of a body's rejecting therapy.

## Stem Cell Therapy in CNS Disorder

Neural stem cell-mediated therapy for rare brain diseases: LSDs and MNDs.

Lysosomal storage diseases (LSDs) are genetically inherited disorders affecting most patients in paediatric age and progressively lead to severe, even lethal, multi-organ dysfunction and brain neurodegeneration. Motor neuron diseases (MNDs) or Amyotrophic Lateral Sclerosis (ALS)- related syndromes are neurodegenerative disorders occurring in the majority of cases sporadically and affect adult middle-aged patients. Despite being divergent in most pathological and physiological hallmarks, both MNDs and LSDs are characterized by tremendous clinical heterogeneity due to poor prognosis and variable onset of the symptoms. Moreover, both LSDs and MNDs are characterized by the concurrence of multiple pathogenetic processes such as the development of inflammatory and excitotoxic environments. Furthermore, pharmacological, enzyme or genetic therapies have proven to be ineffective and no cure is currently available for the neurodegeneration in either LSD or ALS affected patients. Recent studies have identified non-neuronal cell types such as astrocytes and microglia, as being involved in non-cell autonomous effects on MND or LSD progression. These findings have prompted the use of neural stem cells for the replacement of non-neuronal cells rather than neuronal cells which may result in neuroprotection and immunomodulation. The choice of an appropriate tissue source and the establishment of standardized paradigms to culture human neural stem cells (hNSC) will allow their use for future clinical trials on both ALS and LSD affected patients and parallel drug screening studies with novel breakthroughs in the knowledge of neurodegenerative diseases.

## Stem Cell Therapy for Autism

Autism is a spectrum of disorders characterized by marked abnormalities in communication and social interactions.

Two common consistent findings are associated in children with this disorder:

1. Diminished oxygenation in specific areas of the brain and
2. A chronic immunologically mediated inflammatory condition in the gut.

Current investigative therapies for autism attempt to reverse these abnormalities through administration of antibiotics, anti-inflammatory agents, and hyperbaric oxygen. Unfortunately, none of these approaches address the root cause of oxygen deprivation and intestinal inflammation. Mesenchymal stem cells can regulate the immune system. It is thought that they may help to reverse inflammatory conditions and is currently in the final stages of clinical trials for Crohn's disease, a condition resembling the gut inflammation in autistic children. The CD34 stem cells have been shown to induce generation of new blood vessels in tissue that lacks oxygen. It is thought that these stem cells may help increase the blood flow in the temporal region of the brain, an area that is believed to be inadequately oxygenated in autistic children.

Through administration of mesenchymal and CD34 stem cells, patients have shown improvements and good prognosis observed.

The adult stem cells used to treat autism are derived from expanded and fresh human umbilical cord (allogeneic mesenchymal and CD34+). These stem cells are recovered from donated umbilical cords. Before they are approved for treatment all umbilical cord-derived stem cells are screened for viruses and bacteria to International Blood Bank Standards.

Umbilical cord-derived stem cells are ideal for the treatment of autism because they allow our physicians to administer uniform doses and they do not require any stem cell collection from the patient, which for autistic children and their parents can be an arduous process. Because they are collected right after (normal) birth, umbilical cord-derived cells are much more potent than their "older" counterparts like bone marrow-derived cells for instance. These stem cells pose no rejection risk because the body does not recognize them as foreign.

The umbilical cord-derived stem cells are administered intravenously and depending upon the age and physical size of the patient, the stem cells might also be administered intrathecally (into the spinal fluid) by an experienced anesthesiologist. Intrathecal injection allow the stem cells to bypass the blood–brain barrier and migrate throughout the central nervous system.

The autism treatment protocol typically takes 5 days.

## Stem Cell Treatment – Autism: Protocol Outline

### Option 1

- Treatment length : 5 days
- 4 intravenous infusions of expanded/non-expanded donor mesenchymal and CD34+ stem cells.

### Option 2

- Treatment length : 5 days
- 2 intravenous infusions of expanded/non-expanded donor mesenchymal and CD34+ stem cells.
- 2 intrathecal infusions of expanded/non-expanded donor mesenchymal and CD34+ stem cells.

Proper follow-up is an essential part of the autism treatment process. Our primary goal is to ensure that the children are progressing safely. Regular follow-up also enables us to evaluate efficacy and improve our autism treatment protocols based on observed outcomes.

Medical review is done after 1 month, 3 months, 4 months, 1 year and 2 years to monitor the child's progress.

# STROKE AND STEM CELL THERAPY

## Stem Cell Therapy in Respiratory Diseases

Two genetic diseases of the respiratory tract appear amenable for treatment by gene and cell therapy: cystic fibrosis and alpha 1 anti-trypsin deficiency. Cystic fibrosis is caused by one of 1,400 mutations in a single gene, the cystic fibrosis transmembrane conductance regulator (CFTR). Its discovery ignited high hopes for a rapid cure of cystic fibrosis by gene therapy. Various gene transfer vectors that express CFTR have been tested in one or more gene therapy clinical trials. These trials exhibited a good safety profile and focused the scientists on challenges necessary to provide sufficient, long-term CFTR expression in the lung. The progress and status of gene and cell therapies for these respiratory diseases provides a good progress in their treatment. Most patients born with respiratory diseases are not able to breathe as easily as healthy people. The most common genetic respiratory disease is cystic fibrosis. Although cystic fibrosis affects other organs (pancreas, liver, gall bladder, sweat glands, and intestines), lung disease usually causes the most discomfort and is ultimately responsible for the death of the vast majority of CF patients. Cystic fibrosis is caused by mutations in a gene called the cystic fibrosis transmembrane conductance regulator (CFTR). Both copies of the CFTR gene inherited from the father and mother must have mutations to get the disease. Most patients (70%) harbour the same, very common, mutation in CFTR. The large CFTR protein is present in the cells that compose the lung tissues called epithelial cells, which are the cells lining the airways in the lung. The CFTR protein helps epithelial cells regulate the passage of salt and water across the epithelial layer. Therefore, patients with cystic fibrosis have altered concentrations of specific salt molecules (such as sodium and chloride) in their lungs. The observation that people with one normal CFTR gene have normal lung function indicates that expression of at least half the amount of normal CFTR in all relevant cells would reduce or eliminate the symptoms. The discovery of the CFTR gene ignited high hopes for a rapid cure of cystic fibrosis by gene therapy. Various vectors that express CFTR have been tested in one or more clinical trials. Although the recent clinical trials have demonstrated safety in the treated patients, minimal efficacy was evident. Most scientists agree that the ideal vector would efficiently target relevant epithelial cells of the lungs, express sufficient CFTR protein, hide from the immune system, remain invisible to the inflammatory cells, and be sustained in the cells or organ for years. Alternatively, a vector that maintains sufficient expression by repeated administration would be acceptable. Scientists are investigating viral and non-viral vectors to develop the ideal vector with these characteristics. The airways have evolved many ways to protect against the outside environment, and these mechanisms hinder gene transfer to lung epithelial cells. These challenges include airway mucus, specific sugar molecules that block binding of viruses or vectors, and an army of immune cells and inflammatory cells that can intercept any foreign viruses or substances. These same protective mechanisms can hinder the uptake of the gene therapy agent.

Most cases of alpha 1 anti-trypsin deficiency are associated with 1 of the 100 mutations found in the alpha 1 anti-trypsin gene. Two copies of this low producing allele (Z) results in an alpha 1 anti-trypsin protein level ranging from 3.4–7 μmol/L, which is significantly lower than the 40–53 μmol/L level in normal people. This variant of the alpha 1 anti-trypsin protein folds in an abnormal manner and this altered structure prevents its release from liver cells, which can cause liver cell death. The severity of alpha 1 anti-trypsin deficiency is variable. For example, the null allele does not produce any alpha 1 anti-trypsin protein, increases the risk for emphysema, and causes no liver disease as there is no protein to over- accumulate. The more severe causes show symptoms of emphysema, chronic obstructive pulmonary disease (COPD), and occasionally chronic liver disease. People who carry one normal allele and one null allele have no symptoms and express about one third (35%) of the normal level of alpha 1 anti-trypsin.

A combined gene and cell therapy strategy is also being explored for treatment of cystic fibrosis and alpha 1 anti-trypsin deficiency in animal models. The concept of such a therapy is as follows. Basically, the stem cells from the patient would be treated with a gene therapy agent that will allow the expression of a normal copy of the mutated gene. The cells would then be allowed to grow in number and transferred back into the patient. To meet these goals, ongoing scientific research is identifying the relevant stem cells, developing methods for their successful harvest or maturation, trying to develop engineering vectors with the appropriate characteristics to maintain long term expression of this gene, and reduce detection of the corrected cells by the immune and inflammatory cells that could potentially eliminate them. After appropriate testing in tissue and animals, these studies can eventually lead to combined gene and cell therapy strategies for the treatment of cystic fibrosis and alpha 1 anti-trypsin deficiency.

## Spinal Cord Injury and Stem Cell Therapy

After a spinal cord injury, many of the nerve fibres at the injury site lose their insulating layer of myelin. As a result, the fibres are no longer able to properly transmit signals between the brain and the spinal cord contributing to paralysis. Unfortunately, the spinal cord lacks the ability to restore these lost myelin-forming cells after trauma.

Tissue engineering in the spinal cord involves the implantation of scaffold material to guide cell placement and foster cell development. These scaffolds can also be used to deliver stem cells at the site of injury and maximize their regenerative potential.

When the spinal cord is damaged—either accidentally or as the result of a disease (multiple sclerosis, infections, tumours, severe forms of spina bifida, etc.)—it can result in the loss of sensation and mobility and even in complete paralysis.

## Embryonic Stem Cell Therapy Shows Steady Benefits in Rebuilding Infarcted Heart

Using embryonic stem cells, researchers transformed these master cell types into fully functional cardiac cells and transplanted them into damaged regions of the myocardium, where the cells integrated into the infarct and showed rapid and robust improvements, which were stable over an extended period of post-therapy.

The study, entitled "Stable benefit of embryonic stem cell therapy in myocardial infarction", appears in the August 2004 edition of the American Journal of Physiology – Heart and Circulatory physiology, one of 14 peer-reviewed journals published by the American Physiological Society.

Researchers took a murine embryonic stem cell line and engineered a cell clone to express fluorescent proteins so they could easily find the location of the "newly" generated cardiac cells. They tested the carcinogenic capacity of the line and collected cells with high potential for becoming cardiomyocytes. Randomly assigned rats that had been induced with myocardial infarction were injected directly into the damaged heart area either with the embryonic stem cells or were subject to a control of "sham" protocol.

Three weeks after therapy, the cardiac contractile function of both groups was tested by echocardiography. Not only was the stem cell-treated group's left ventricular pumping significantly stronger than the sham-treated group, but the heart beat of the stem cell group reacted favourably to "stress test", whereas the sham group showed no significant response.

Like all the other parameters, these improvements were maintained over the three-month length of the study.

On pathology, further investigation showed the extent of the positive influence of the stem-cell therapy. First, cardiac cells stayed in the heart, and did not spread to the brain, kidney or liver. Microscopy showed that the stem cell-derived cardiomyocytes took on the distinctive striations indicating proper development of contractile apparatus. Stem cell- treated hearts also showed normal cardiac ultrastructure, in contrast to cellular infarcted areas of sham-treated hearts.

Finally, the stem cell-treated hearts showed that the wall or muscle had been "rebuilt" compared with the sham-treated hearts which remained " eaten-up", with a decayed thin look, including formation of aneurysms, associated with a post-heart attack to rupture. Conversely, the stem cell-treated hearts showed no evidence of graft rejection, electrical and/or structural abnormality, sudden cardiac death or tumour formation.

The researchers conclude that "embryonic stem cells, through differentiation within the host myocardium, can contribute to a stable beneficial outcome on contractile function and ventricular remodelling in the infarcted heart."

"The stable benefit of embryonic stem cell therapy on myocardial structure and function in this experimental model supports the potential for stem cell-based reparative treatment of myocardial infarction. By regenerating diseased myocardium and promoting cardiac repair, embryonic stem cells provide a unique therapeutic modality that has the potential to reduce the morbidity and mortality of this prevalent heart disease."

The researchers also noted that issues to be resolved include mechanisms of action, finding the optimum window for therapy, and determining what the long-term effect of such therapy will be.

## General Scientific Discovery

Stem cell research is also useful for learning about human development. Cancer and other birth defects are also affected by abnormal cell division and differentiation. New therapies for diseases may be developed.

Another reason why stem cell research is being pursued is to develop new drugs. Scientists could measure a drug's effect on healthy, normal tissue by testing the drug on tissue grown from stem cells rather than testing the drug on human volunteers.

## STEM CELL CONTROVERSY

The debates surrounding stem cell research primarily are driven by methods concerning embryonic stem cell research. It was only in 1998 that researchers from the University of Wisconsin-Madison extracted the first human embryonic stem cells that were able to be kept alive in the laboratory. The main critique of this research is that it required the destruction of a human blastocyst. That is, a fertilized egg was not given the chance to grow into a fully-developed human.

### When does life begin?

Many assert that life begins at conception, when the egg is fertilized. It is often argued that the embryo deserves the same status as any other fully grown human. Therefore, destroying it (removing the blastocyst to extract stem cells) is akin to murder. Others, in contrast, have identified different points in gestational development that mark the beginning of life—after the development of certain organs or after a certain time period.

### Chimeras

A chimera is an organism that has both human and animal cells or tissues. Often in stem cell research, human cells are inserted into animals (like mice or rats) and allowed to develop. This creates the opportunity for researchers to see what happens when stem cells are implanted.

### Legal Issues

The stem cell debate has risen to the highest level of courts in several countries. Production of embryonic stem cell lines is illegal in Austria, Denmark, France, Germany, and Ireland, but permitted in Finland, Greece, the Netherlands, Sweden, and the UK. In the United States, it is not illegal to work with or create embryonic stem cells lines. However, the debate in the US is about funding, and it is in fact illegal for federal funds to be used to research stem cell lines that were created after August 2001.

## MEASURES TO INCREASE STEM CELL SURVIVAL

Even if stem cells have an intrinsic capacity to create new tissue, they are unlikely to do so if the recipient environment is not conducive to regeneration.

### Step-I

### Increasing Stem Cell Survival by Preconditioning

Survival of injected stem cells is influenced by the recipient environment. For instance,

i. Laflamme *et al.* injected human mesenchymal stem cells (hMSCs) into areas of rat peri-infarct and normal myocardium. After 28 days, 90% of labelled hMSCs were found in the normal myocardium, whereas in the infarcted myocardium, only 18% of injected human cells were detectable. Ischemia is thought to create an inhospitable environment by inducing local expression of inflammatory cytokines that promote cell apoptosis.

ii. Alternatively cell survival can be enhanced by inhibition of local inflammation. Adenosine in addition to its well known cardiac effects is a potent natural anti-inflammatory agent. Patel *et al.* showed that in canine myocardial infarction (MI), an adenosine agonist reduced infarct size by 16% when given at the time of coronary reperfusion.

In continuation of attempts to improve cell survival, Pons *et al.* showed that injection of vascular endothelial growth factor to an MI heart increased MSCs survival by 2.9 fold in ischemic myocardium and improved therapeutic effects of MSCs for treatment of myocardial infarction. The timing of therapies to reduce ischemia – induced cell apoptosis appears to be critical. For instance, when the large randomized, placebo-controlled

AMISTAD-2 trial 1 was subjected to adenosine infusion as an adjuvant to thrombolytic reperfusion therapy, within the first 3.2 hours onset of evolving MI, infarct size was reduced 14% more compared with therapy after that cut-point.

### Step-II

## Promotion of Cell Growth and Differentiation

Many factors that promote growth and differentiation are also anti-apoptotic. For instance, growth factor G-CSF1 not only substantially inhibits apoptosis but also promotes cell proliferation. Bartunex *et al.*, pretreated MSCs with fibroblast growth factor (FGF) to induce a 10% increase in regional wall thickening of the infarcted territory in the chronic dog Myocardial Infarction model. Lipopolysaccharide is also one of the other pro-proliferative factors that have been tested.

## METHODS TO INCREASE CELL SURVIVAL?

Anything that improves general health, immune status and circulation will improve the effectiveness of stem cell treatment.

1. **Exercise** Exercise has been shown to increase the number of stem cells in the circulation. A vigorous circulation is essential for cell growth.

2. **Diet** Everything that we eat will have an impact on cell survival. Fruits and vegetables rich in antioxidants and phytonutrients will actively support an optimal environment for cell growth. Saturated fats and high carbohydrates have been shown to be inflammatory and have a negative effect.

3. **Supplement** There are many supplements that look very promising. These include:

   i. Resveratrol.

   ii. Carnatine

   iii. Colostrum

iv. Fish Oil

v. Vitamins, minerals and antioxidants.

# STEM CELL RESEARCH

Till date, stem cell research is a controversial issue all over the world. While some argue about its benefits, others argue about the potential threat it poses to humanity. Here, we try and understand the basics of stem cell research and its pros and cons.

Any discussion on stem cell research must first begin by understanding what stem cells exactly are. Stem cells are basically a particular type of cells that are primitive or immature, and that can be transformed into specialized cells that possess the ability and the functions of the human body. There are two main sources of stem cells namely, embryonic stem cell (embryos that have been formed in the blastocyst phase in the process of embryological development) and adult stem cells (those obtained from adult tissues). Given the abilities of stem cells to develop into independent cells, a lot of research has been done into these varieties and types of stem cells. Though there are two sources of stem cells, greater attention has been paid to embryonic stem cell research. Hence, an attempt has been made to understand what stem cell research is and the ethics of this kind of a research, made various embryonic stem cell research pros and cons.

## What is Stem Cell Research?

The process of stem cell research is one that extracts stem cells and transforms them into any or all of the 220 types of cells found in the human body. These cells even include blood and brain cells. As these cells possess the ability to be converted into cells of the human body, researchers believe that these can be used to treat a variety of health conditions faced by human beings, such as heart disease, cancer, diabetes, and conditions of the brain such as Parkinson's disease and Alzheimer's disease. This is one of the biggest stem cell research benefits.

The reason embryonic stem cells are preferred over adult stem cells is that they are easier to extract. Though adult tissue can be obtained with little or no trouble, it has been found that embryonic stem cells possess greater potential in discovering the power of stem cell therapy. Embryonic stem cells are created to carry out stem cell research, or the excess available from the process of *in-vitro* fertilization (IVF) may be used for this purpose, though the latter case is slightly rare. As is the case with the development of any new scientific technology, numerous advantages and disadvantages of stem cell research have been discovered. A huge discussion has emerged about the ethics behind carrying out embryonic stem cell research.

## Stem Cell Research: Ethics and Issues

From a detailed look into the various stem cell research facts, two important issues regarding the ethics of stem cell research have emerged.

- One of the biggest threats posed by stem cell research is the possibility of human cloning. As mentioned earlier, stem cells possess the ability to be differentiated into the cells possessed by human beings. If stem cells can be modified to clone human cells, it poses a big threat to humanity.
- Another big controversy in stem cell research is that embryonic stem cell research marks the end of potential human life. According to the propagators of this side of the debate, an embryo marks the beginning of human life. When embryos are isolated in the blastocyst phase of embryological development, it results in the death of the embryo. This they believe, is equivalent to murder and such an act should be punishable by law.

Though stem cell research has a lot of positive things to offer, it has a negative side too. As such, the various pros and cons of stem cell research have to be carefully considered to be able to identify its scientific potential and its ethical positions. If carried out in an appropriate manner, stem cell research and therapy can prove to be a big boon to human life. If fallen in the wrong hands, this process can be moulded to create disastrous consequences. It is essential to have specific guidelines and judicial laws governing stem cell research, only after which can this effort prove useful to humanity.

As we know, stem cells are special kind of cells that have the ability to divide indefinitely and have the potential to give rise to specialized cells (that is, any cell of the body). Stating it in scientific terminology, the defining properties of stem cells are:

- They are not terminally differentiated but determined.
- They can divide indefinitely.
- When they divide, each daughter cell has a choice: it can either remain a stem cell or take on a course leading to terminal differentiation.

The job of stem cell is not carry out differentiation function, but rather to produce cells that will differentiate. Consequently, stem cells have a not-descript appearance, making them hard to identify. But this is not to say that the cells are alike. Although not terminally differentiated, they are nevertheless determined.

Adult-derived stem cell therapies will complement, but they cannot replace, therapies that may be obtained from embryonic stem cells. Still, they do have some advantages. For example, adult stem cells offer the opportunity to utilize small samples of adult tissues, to obtain an initial culture of a patient's own cells for expansion and subsequent implantation in the same person (that is called an autologous transplant). This process avoids immune rejection by the recipient and also protects the patients from viral, bacterial or other contamination from another individual (donor) in case of allogenic transplant. Autologous and allogenic transplants of hematopoietic stem cells that are isolated from mobilized peripheral blood or from bone marrow by positive selection with antibiotics are in clinical use. Additionally, they normally differentiate into a narrow set of cell types, directing them to a desired fate is easier.

One major disadvantage is that culturing adult stem cells *in-vitro* is very difficult and has not been possible from some types. Also they have a very short life, when cultured *in-vitro*, as compared to embryonic cells.

## Advantages of Embryonic Stem Cell Research

Embryonic stem cell research is a controversial topic amongst many group, however, its advantages are numerous. Embryonic stem cells are the basic building block of human life, encoded with all the information on human genetics. Diseases, defects and injuries can all be studied using embryonic stem cells.

1. ***Features*** The number one advantage of embryonic stem cell research is that they can be manipulated and grown into any type of cell in the human body. Unlike adult stem cells embryonic cells contain the genetic information for every cell type. This means that scientists can use cell to grow nerve, blood tissue or other cells. The advantage of this is that scientists can see how and why a certain cell develops. This will aid them in understanding the difference between different cells and their full function within development and growth.

2. ***Potential*** By understanding cell development using embryonic stem cells, doctors and scientists will be able to prevent many birth defects. Different forms of defects occur at varying stages of development within the womb. Studies on embryonic stem cells will allow them to find the exact point at which certain cells mutate and a birth defect is formed. In theory they will be able to prevent this from occurring either by stimulating or blocking certain enzymes while in the womb. Approximately 3 percent of babies are born with birth defects in the United States according to the University of Virginia. Scientists believe they can limit at least 1 percent of these defects using embryonic stem cell research.

3. ***Benefits*** Because embryos are kept in a permanent state of early development, they are ideal candidates for drug and medical testing. While they are still alive, without a host body in which they can develop, they will never become fully human. However, since they are alive, testing can be conducted in a variety of fashions. A scientist will be able to infect them with a disease and attempt to test various treatments. Advances in the research of cancer, spinal cord injuries, diabetes, Parkinson's, Alzheimer's and schizophrenia can all benefit

from this research. In addition, no human or animal testing needs to be conducted, which will save the lives of both.

4. ***Significance*** Early testing on embryonic stem cells has shown to be beneficial in understanding the aging process. Beyond the research about early human development, scientists can identify the specific enzymes that cause the aging process to work. Certain proteins are created in our cells which cause humans to age at a specified rate. These enzymes can be blocked or removed to allow for a slower aging process. The possibility even exists that the entire aging process itself could be stopped. This will mean longer and more productive lives for many people. Every year, thousands of people suffer injuries from the aging of their bodies, whether it is organs that fail or damage sustained from a fall. Without the extreme process of aging, many of these occurrences would be eradicated.

5. ***Considerations*** One of the greatest advantages of embryonic stem cell research is the possibilities associated to cloning. Embryonic stem cells are the ideal candidates for cloning. They contain information on every cell in the body and they can be manipulated in nearly any way imaginable. Scientists believe they can grow limbs and organs that could be transplanted into humans. These could be exact DNA matches which would make the transplant process more efficient and have a greater chance of succeeding. Research also shows that embryonic stem cells have the ability to regenerate, opening up the opportunities that an organ or tissue already on a human body could be regrown. Amphibians and reptiles have the ability to regenerate body parts, logically the genetic information should be in human DNA as well.

# 19

# Stem Cells and Renal System

The kidney comprises a complex interacting network of tubules formed by epithelial cells, capillaries, arterioles, interstitium, lymphatic vessels and neuronal cells. In mammals, nephrogenesis stops at birth, while in some lower vertebrates like sharks new glomeruli continue to form throughout life in the nephrogenic areas. In humans, after ischemic insult to the tubules, regeneration of tubular epithelial cells is possible but the specialized renal cells like the podocytes are terminally differentiated and cannot regenerate.

Recent studies point to the existence of endogenous renal adult stem cells, which can exist for long periods without dividing. These slow cycling cells have been found in the interstitium of the renal papilla. After a period of ischemia, these cells start dividing and migrate to the site of injury. Another approach to identify adult stem cells is by the side population phenotype. These side population (SP) stem like cells, which are not derived from the bone marrow, contribute to tissue regeneration without trans-differentiating into damaged cells. Bone marrow stromal cells [also called mesenchymal stem cells (MSCs)] may play a role in renal regeneration. These cells which are CD45 negative make up less than 0.01% of the total bone marrow cell population. They probably protect the kidney via paracrine and or endocrine effects rather than through direct engraftment in renal tubules.

# STEM CELL THERAPY IN RENAL DISEASES

Acute renal failure /acute tubular necrosis pharmacologic regimens that mobilize MCSs could aid repair. However this is fraught with danger as inflammatory cells are also mobilized. Another approach is to amplify MSCs in the laboratory and then re-infuse them into the recipients at time of injury. This is difficult to practice in clinical situations. A third approach would be to identify the cytokines /growth factors secreted by the MSCs which are responsible for the renal reparative action, and administer them directly.

Activation of SP cells by Trichostatin A has been shown to prevent progression of chronic renal failure (CRF) in animal models.

Genetic disorders, stem cell therapy has also been tried in mouse models of Fabry disease and Alport syndrome.

## Stem Cell Therapy for Urinary Stress Incontinence

The increasing incidence in urinary incontinence (UI) with advancing age is directly correlated to spontaneous apoptosis of the muscles cells of the rhabdosphincter, the striated urethral sphincter. Age-dependent reduction in the number of striated muscle cells is now held the highest ranking cause of urinary stress incontinence in elderly women and men. The technical availability of autologous satellite cells obtained from skeletal muscle enabled the immediate application of muscle progenitor cells in the treatment of UI by means of tissue engineering techniques in patients of any age, including elderly.

This review focuses on the theoretical assumptions and experimental findings on muscle derived stem cells (MDSC) justifying their straightforward application in urology. Only quite recently, the regenerative function of mammalian tissues has become understood as ubiquitous. Not only rapidly renewing tissues like skin, bone marrow or blood were found to possess a stem cell reserve, but also, apparently quiescent tissues like nerve tissue or heart muscle disclosed during cellular stress or tissue loss an inherent renewing potential ( Hanashima et al., 2004). Tissue renewal depends on the persistence of so called adult stem cells or precursors of mature cells.

Moreover not only acute processes like injury or intoxication activate stem cell directed tissue regeneration. Cellular aging, cell death (apoptosis and necrosis) and renewal continue throughout life and through all tissues. Imbalance between cell death and renewal are now thought to occur during aging (when apoptosis supersedes renewing capacity) and during malignant transformation (when regular cell death is hindered by apoptosis inhibition) (Hanahan and Weinberg, 2000). According to this rationale, the higher prevalence of UI among elderly persons of both sexes can be interpreted as a symptom of increasingly poor and eventually failing tissue regeneration in the vesicourethral apparatus.

A direct correlation between age and the number of apoptotic cells in the rhabdomyosphincter on the one hand, and impaired detrusor contractility and weak closure pressure caused by structural deficits of the rhabdosphincter on the other hand, suggest that stress incontinence is predominantly associated with a dysfunction of the striated urethral sphincter (Frauscher et al, 1998 Strasser et al, 1999). Yiou and co-workers have shown that the adult rhabdosphincter in mice regenerates after an injury by means of intrinsic satellite cells (Yiou et al., 2003). This finding is interesting per se because rhabdoshpincter and striated skeletal muscles have a different embryological origin. Muscle cells of the external sphincter are thought to arise via transdifferentiation of urethral smooth muscle cells and thus would not normally be expected to generate myogenic satellite cells whereas striated skeletal muscle develops from somatically derived myoblasts (Perry and Rudniciki, 2000).

The sources for sphincter-associated satellite cells have not been finally elucidated. Preclinical studies, however, demonstrated that satellite cell-mediated sphincter regeneration can be achieved *in vivo*. The regenerative capacity of the rhabdosphincter declines in elderly, poor dietary or training situation and the idea of an at least partial compensation of this loss by implantation of heterotopic autologous satellite cells was suggestive.

Feasibility of this concept was demonstrated in animal experiments (in rats by Chancellor et al., 2000, Cannon et al 2003, in mice by Yokoyama et al., 2001, in pigs by Strasser et al., 2003) Experiments conducted with betagalactosidase tranfected cells allowed visualization of the injected cells *in vivo*, their persistence in the sphincter structure, differentiation into myotubes and histological orientation *in vivo* (see also Fig 1A and B). Fascination about stem cells largely lays in their developmental plasticity. Differentiation of totipotent embryonic stem cells leads to progressively restricted, committed cell lineages. According to the current dogma, adult stem cells are confined to become one particular adult cell type, i.e., satellite cells will develop into myoblast and myotubes only, but not into other cell types. Restricted plasticity of satellite cells may hold lower fascination, their restrictedness argues, however for more safety in clinical applications. Apparently, the injected myoblasts are able to substitute cellular losses in defective structures (Irintchev et al., 2002) but do not multiply beyond the required repair needs, which might be explained in analogy to normal haematopoiesis by the interplay between cytokines and transcription factors.

The risk of tumour formation—the worst case scenario for late consequences of tissue engineering and cellular therapy—cannot be finally ruled out until we know the results of long term studies. By now however there are no indications that the normal, non- transformed autologous myoblasts behave irregularly *in vivo* (Irintchev al., 2002). Extensive literature searches did not bring about any hints on granulomas, chronic inflammation, tumours or incompatibility reactions consequent to injection/implantation of autologous myoblasts, neither in animal experiments nor in clinical trials.

Experimental and clinical data on the fate of heterotopic (though autologous) stem cells are scarce. The expectations in both the efficacy and safety are particularly high in indications such as UI.

## Experimental Results

Recent sonographic and experimental data suggest that the urethral closure pressure is predominantly provided by the rhabdosphincter (Koelbl et al., 2001; Strasser et al. 1998). First step towards a new therapy of sphincter insufficiency were reported by Chancellor et.al.(2001) Cannon et.al (2003) and Yokoyama et.al (2001). Using rodent models these authors demonstrated that intraurethral injections of (allogenic) immortal myoblasts that had been developed from skeletal muscle biopsies are feasible and accomplish regeneration of the sphincter. Chancellor et.al (2001) and Cannon et al.

(2003) focused on the implantation of allogenic rather than autologous and survival of these allogenic MDSCs(Myeloid-derived suppressor cells) in the rhabdosphincter/urethra as a perhaps immunologically privileged site.

Experimental work on rhabdosphincter repair with MDSCs proceeded from rodents to a large animal model. A feasibility study in pigs had to proof, whether porcine autologous MDSCs and fibroblasts can be harvested, propagated in culture, injected under sonographic control into the pig urethra, and visually recovered *in vivo* (Strasser et.al., 2003)

Skeletal muscle samples were obtained from eight pigs. Each biopsy was enzymatically dissociated using a modified cell dispersion technique to grow pure monoclonal cultures of myoblasts and fibroblasts. Single myoblasts in suspension were manually collected with a micropipette under microscopic control. Desmin served as a marker to identify clones of myoblasts. Pure culture of myoblasts and fibroblasts were maintained in a proliferative status for several months. Before implantation, the cells underwent fluorescence labelling to allow their subsequent recovery *in vivo* following injection and to assess the degree of integration into the urethra and rhabdosphincter. Labelled cell samples were injected into the urethra under direct sonographic control employing a transurethral ultrasound probe.

Transurethral injection of myoblasts under direct sonographic control proved to be an easy, quick and minimally invasive procedure. Histologic examination of urethral specimens revealed that the injected cells survived for at least 10 weeks. Fluorescent images suggest that myoblasts formed new myofibres (and /or) alternatively, fused with the pre-existent myofibres.

Observation of prolonged survival and integration of myoblasts in the sphincter area and postoperatively, increased urethral closure pressure (Strasser et al., 2003) and facilitated the rapid progression from the preclinical to clinical testing stage of incontinence treatment by means of adult stem cells.

## Present Therapies

Current treatment modalities of urinary stress incontinence produce high initial success rates, but are considerably invasive. For example, retropubic suspension can be expected to generate an immediate mean cure rate of 79% and an improvement rate of 90%. Sling procedures, using autologous or synthetic materials, produce a cure rate of approximately 80% and an improvement rate of 90% one year after surgery. Injection of bulking agents are regarded a less invasive method to treat stress incontinence, but the improvement is often short-lived and overall success rates are poor.

## First Results of Clinical Studies

*In vitro* studies, results on implantation of MDSCs into the porcine rhabdosphincter, and careful safety considerations justified the primary application of the MDSC implantation technique in patients with urinary stress incontinence. Meanwhile, the growing number of successfully treated patients serves as proof of a concept, which is per se convincing in its simplicity; the patient is both donor and recipient of his own cells, while the site of donation and implantation are different, and thus the propagated cellular material can be termed an autologous, heterotopic, targeted donation.

Between September and December 2002, five female patients suffering from urinary stress incontinence (age: 36–78 years ) were enrolled to participate in a phase I/II open, baseline-controlled, monocentric study of ultrasound guided transurethral injection of autologous muscle stem cells (Strasser et al. 2004)

Before injection, skeletal muscle biopsies were taken from the left arm to obtain cultures from autologous myoblasts and fibroblasts. Incontinence score, urodynamic, electromyographic and laboratory parameters, changes in quality of life as well as morphology and function of urethra and rhabdosphincter were evaluated at baseline and postoperatively upto 1 year after therapy. Transurethral ultrasound and three-dimensional ultrasound were used to investigate lower urinary tract. While myoblasts were directly injected into the rhabdosphincter to reconstruct the muscle, fibroblasts were mixed with collagen and injected into the urethral submucosa to treat atrophies of the mucosa. After the injection, i.e., beginning from day 1 postoperatively, the incontinence score of the patients, improved, until they became completely continent within the first weeks. One year after injection of autologous stem cells, the five formerly incontinent patients had remained symptom-free and their life quality had dramatically improved. Ultrasound visualization revealed that thickness of urethra and rhabdosphincter as well as activity and contractility of the rhabdosphincter had significantly increased as compared to baseline (Strasser et al., 2004 ). At this stage, we can only speculate if myoblasts form new fibres or, alternatively, strengthen sphincter contrac-tility by fusing with existing muscle fibres. The extent of integration and regular participation of new cells in the sphincter function are currently under intense research. Furthermore, engrafted myoblasts may affect the sphincter function through paracrine effects mediated by release of growth factors such as HGF (Menasche et.al., 2003). A strengthening of the sphincter musculature after myoblast implantation would however provide the best explanation for the dramatically improved condition of the five patients persisting by now for more than 12 months (Strasser et al., 2004). However, the exact mechanism responsible for the clinical improvement is still elusive, although the amount of collagen used as carrier medium during implantation is minimal (only upto 2.5ml). We cannot exclude that it exerts a transient bulking effect during the first post-implant days. Likewise, the role of fibroblasts needs careful investigation. Hence the impact of three components, myoblasts, fibroblasts and collagen, requires assessment in a well-designed prospective, clinical trial with long term follow-up.

## CORD BLOOD BANKING

Cord blood banking is a procedure in which cord blood—a rich source of stem cells—is taken from a new born's umbilical cord after delivery and preserved for possible future use in a stem cell transplant. Collecting a baby's cord blood poses few, if any, risks to either mother or baby.

Cord blood transplants from unrelated donors can be used to treat many conditions, ranging from leukaemia and various metabolic conditions to cerebral palsy.

If one is interested in preserving their baby's cord blood for possible personal use, he/she might consider using a private cord blood banking facility.

Top 5 reasons for the consideration of cord blood banking.

- 1 in 200 individuals may need a haematopoietic stem cells (HSCs) transplant by the age of 70 according to a scientific paper published in 2008.
- It is once-in-a-life time chance to collect cord blood—a readily available source when needed in the future.
- 60% higher chance of locating a match versus bone marrow in a family.
- Research showed that patients have a lower chance of complications in transplants when they receive stem cells from a relative.
- Some of the most common cancers are treatable with a stem cell transplant. For instance, lymphoma and leukaemia are two of the top 10 common cancers for adults and among the top five most common childhood cancers.

## Cord Blood – Source of Hematopoietic Stem Cells

Human cord blood has numerous advantages over other stem cell sources which are listed below.

- Quantity of hematopoietic stem cells are high in cord blood.
- Immune rejection of hematopoietic stem cells collected from cord blood does not occur in the recipient. This is associated with fewer complication related to the transplant.
- Cord blood donation is a painless procedure and does not hinder the birth process in any way.

## Technique of Collection

About 250 ml of cord blood is collected with anticoagulant and stored in liquid nitrogen.

The total nucleated cell count collected from each cord blood donation is 0.5 million to 5 billion cells.

## Problems Associated with the Use of Cord Blood Stem Cells

- Wastage of a majority of the collected cord blood samples.
- The infrastructural need for storage of cord blood is very high.
- The potential of cord blood stem cells in the treatment of human diseases is not yet fully understood.
- Molecular and immunological analysis of cord blood is difficult.
- There is an increased risk of transmitting disorders to the recipient.
- The amount of cord blood collected from the donor is highly variable and is largely dependent on the skill of the operating room personnel.
- Re-collection of the cord blood sample from the same donor is not possible.
- Risk of maternal blood contaminations.

## MEDICAL CONTROVERSY ON CORD BLOOD BANKING

Umbilical cord blood is well recognized to be useful for treating hematopoietic and genetic disorders, but even then some controversy surrounds the collection and storage of umbilical cord blood by private banks for the baby's use. Only a small percentage of babies estimated between 1 in 1,000 and 1 in 2,00,000 ever use the umbilical cord blood that is stored. The American Academy of Paediatrics, 2007 Policy statement on Cord Blood Banking states that "Physicians should be aware of the unsubstantiated claims of private cord blood banks made to future parents that promise to insure infants or family members against serious illnesses in the future by use of the stem cells contained in cord blood;"

Cord blood is stored by both public and private cord blood banks. Public cord blood banks store cord blood for the benefit of the general public, and most U.S banks coordinate matching cord blood to patients through the Nations Marrow Donor Program (NMDP). Private cord blood banks are usually for profit organizations that store cord blood for the exclusive use of the donor or donor's relatives.

Public cord blood banking is supported by the medical community. However, private cord blood banking is generally not recommended unless there is family history of specific genetic diseases.

New parents have the option of storing their new born cord blood at a private cord blood bank or donating it to a public cord blood bank. The cost of private cord blood banking is approximately $2000 for collection and approximately $125 per year for storage, as of 2007. Donation to a public cord blood bank is not possible everywhere. But availability is increasing. Several local cord blood banks across the United States are now accepting donations from within their own states. The cord blood bank will not charge the donor for the donation.

After the first sibling–donor cord blood transplant was performed in 1988, the National Institute of Health (NIH) awarded a grant to Dr. Pablo Rubinsein to develop the world's first cord blood program at the New York Blood Centre (NYBC) in order to establish the inventory of non-embryonal stem cell units necessary to provide unrelated, matched grafts for patients.

In 2005, University of Toronto researcher Peter Zandstra developed a method to increase the yield of cord blood stem cells to enable their use in treating adults as well as children.

***The NHS (National Health Service, UK) Cord Blood Bank*** The NHS Cord Blood Bank was set up in 1996 to collect, process, store and supply cord blood. It is a public cord blood bank and part of the NHS.

- The first unrelated transplant (the patient was not related to the donor) occurred in February 1998.
- Cord blood collected for the NHS Cord Blood Bank is collected by specially trained and dedicated staff. It is collected from the following hospitals—Barnet General, Northwick Park, Watford, Luton and Dunstable and St. George's in Tooting.

- Trained NHS Cord Blood Bank staff contact mothers who have registered to donate before they have their baby to ask simple lifestyle questions and to obtain verbal consent for the collection testing and use of the donation.
- The NHS Cord Blood Bank has issued cord blood units to 22 countries. This is a reciprocal arrangement internally enabling patients worldwide to get the best treatment.
- The NHS Cord Blood Bank is accredited by Net Cord–FACT and licensed by the Human Tissue.

While the American Academy of Paediatrics discourages private banking except in the case of existing medical need, it also believes that information about the potential benefits and limitations of cord blood banking and transplantation should be provided so that parents can make an informed decision. Cord blood education is also supported by legislators at the federal and state levels. In 2005, the National Academy of Science published an Institute of Medicine (IoM) report titled, "Establishing a National Cord Blood Stem Cell Bank Program". The IoM report recommended that expectant parents be given a balanced perspective on their options for cord blood banking. In response to their constituents, state legislators across the country are introducing legislation intended to help inform physicians and expectant parents on the options for donating, discarding or banking life saving newborn stem cells.

## Controversy

While there is general support in the medical community for public banking of cord blood, the question of private banking has raised objections from many governments and non-profit organizations. In March 2008, a paper was published by Nietfeld et al. in the journal Biology of Blood and Marrow Transplantation which computed the lifetime probability (up to age 70 ) that an individual in the US would undergo a stem cell transplant. The likelihood of an autologous transplant using your own stem cells is 1 in 435, the likelihood of an allogeneic transplant form a matched donor (such as a sibling) is 1 in 400 and the net likelihood of any type of stem cell transplant is 1 in 217.

In March 2004 the European Union group on Ethics (EGE) has issued Opinion No.19 titled Ethical Aspects of Umbilical Cord Blood Banking. The EGE concluded that the legitimacy of commercial cord blood banks for autologous use should be questioned as they sell a service, which has presently, no real use regarding therapeutic options. Thus they promise more than they can deliver. The activities of such banks raise serious ethical criticisms. However in the final section 1.27 of their Opinion, the EGE admits that if in the future regenerative medicine developed in such a way that using autologous stem cells became possible, then the fact to have one's own cord blood being stored at birth could increase the chance of having access to new therapies.

In May 2006 the World Marrow Donor Association (WMDA) Policy Statement for the Utility of Autologous or Family Cord Blood Unit Storage stated that:

1. The use of autologous cord blood cells for the treatment of childhood leukaemia is contra-indicated because pre-leukemic cells are present at birth. Autologous cord blood carries the same genetic defects as the donor and should not be used to treat genetic diseases.

2. There is at present no known protocol where autologous cord blood stem cells are used in therapy.

3. If autologous stem cell therapies should become reality in the future, these protocols will probably rely on easily accessible stem cells.

## EMERGING STEM CELL APPLICATIONS (FOR BRAIN INJURY, CEREBRAL PALSY, TYPE 1 DIABETES, HEART DISEASE)

As of spring 2008, there are several known instances where autologous use of cord blood is indicated.

1. Whereas the WMDA cautioned against autologous transplant for disease with a genetic signature, there are paediatric cancers

(e.g., neuroblastoma) and acquired conditions (e.g., aplastic anemia) which can be treated by autologous transplant. There has even been one autologous transplant for leukaemia.

2. Type 1 diabetes, also known as juvenile diabetes has been shown to improve if treated shortly after onset with an infusion of autologous cord blood. The American Diabetes Association reports that 1 in 7000 children is diagnosed each year with Type 1 diabetes, and 1 in 600 children are living with it.

3. A phase I clinical trial is under way at Duke University to investigate whether cerebral palsy and other forms of paediatric brain injury may respond to infusions of autologous cord blood. The Brain Injury Association of America estimates that the prevalence of cerebral palsy is about 1 in 300 among children up to age 10.

# 20

# Stem Cells Therapy for Diabetes

Embryonic stem cells (ESC) can be differentiated into insulin-producing cells by manipulating culture conditions. ESC were stimulated to differentiate towards a cell like phenotype. The addition of phosphoinositide kinase inhibitors promoted differentiation of larger numbers of ESC towards functional cells. With manipulation of culture conditions and use of pax-4 or pdx-1, transcription factor associated with cell lineage yield promising results. ESC differentiation protocols yield cells that produce insulin or cells that absorb insulin from the medium. These differentiated cells must actively synthesize and secrete insulin. Functioning

molecular components of regulated secretion of insulin and insulin-containing vesicles are additional features that indicate a β-cell phenotype.

Transplantation of ESC derived insulin producing cells reverses diabetes in rodents, indicating that these cells do synthesise and release insulin. The early and uncontrolled introduction of transcription factor into ESC during *in vitro* differentiation might not yield the desired results. Regulated expression of introduced transcription factors that can be turned on during *in vitro* differentiation of ESC might be more successful. ESC genetically selected for insulin expression and injected into diabetic rats, improve glucose control.

Human ESC produce insulin under different culture conditions. Techniques that do not require murine feeder cells have been developed, allowing for single species propagation of ESC and avoiding possible zoonotic infection of cells intended for clinical use. Problems in control of differentiation and teratoma formation from ESC derived insulin producing cells remains to be overcome. Existing ESC lines are not assumed to be identical or ideal for generating islets or cells. Hence additional ESC lines continue to be generated. Ethical concerns about the use of ESC need to be addressed and resolved in the face of this powerful technology.

## STEM CELLS DERIVED FROM HAEMOPOIETIC ORGANS

Bone marrow harbours cells that can become parenchymal cells after entering the liver, intestine, skin, lung, skeletal muscle, heart muscle, and central nervous system. In rodent models and in human recipients of marrow or organ transplantation in rodents, haemopoietic organs harbour cells that can also differentiate into functional pancreatic endocrine cells 1–2 months after bone marrow transplantation. Donor-derived cells are found in pancreatic islets of recipient mice. These cells express insulin and genetic markers of cells. In culture, the cells secrete insulin in response to glucose, and show intracellular calcium fluctuations similar to normal cells. However, only about 1–3% of the islet cells originate from the transplanted marrow. A marrow derived cell type with pluri- potential capacity to transdifferentiate into various phenotypes has been described. This or a similar cell type might be able to differentiate into pancreatic cells.

Although these studies show the possibility of bone marrow transplantation as a therapeutic approach for cell replacement, the immunological destruction of newly regenerated cells in type 1 diabetes remains a problem. Bone marrow transplantation induces microchimerism in non-obese diabetic (NOD) mice, an autoimmune model of type 1 diabetes, transplanted with marrow. Before development of autoimmune diabetes, chimerism prevents diabetes mellitus presumably by mechanisms involving donor immune regulatory cells that prevent the host cells from becoming auto- reactive against cells. By contrast, in NOD mice that are already diabetic, induction of chimerism by sublethal or lethal irradiation followed by marrow transplantation does not result in recovery from diabetes. However, when these diabetic transplant recipients are kept normoglycaemic by insulin therapy, they ultimately recover from diabetes. Pancreatic tissue showed increased proliferative activity and regeneration of cells. Thus marrow transplantation to induce immunological control plus maintenance of normoglycaemia allowed local or progenitor cells to proliferate as an adaptive response. Transplantation of mesenchymal cells from the spleen combined with complete Freund's adjuvant let to reversal of diabetes accompanied by regeneration of insulin producing islets. The transplanted splenic mesenchymal cells differentiate into cells. Thus splenic mesenchymal cells transplanted under certain conditions seem not only to keep immune destruction of islets in check, but also can transdifferentiate into pancreatic cells. The relative functions seemed to arise from different subpopulations.

Bone marrow cells can differentiate *in vitro* under controlled conditions into insulin-expressing cells. Such cells transplanted under the kidney capsule of diabetic rodents, correct glucose. Removal of the grafted kidney returned the animals to a diabetic state.

Cell fusion has been suggested as a mechanism of apparent adaption of bone-marrow-derived cells into an extramedullary phenotype. Studies involving pancreatic endocrine-cell differentiation from haemopoietic-organ derivatives largely rule out cell-fusion events as a mechanism of transdifferentiation. Bone-marrow-derived extramedullary parenchymal tissue is not always found.

Some groups find little if any transdifferentiation or bone-marrow derivation of intra-islet endothelial cells in diabetic mice.

The last finding was made recently by Andreas Lechner and colleagues. One group reports the generation of insulin-producing cells in liver, adipose tissue, spleen, and bone marrow in rodent models of diabetes mellitus.

Bone-marrow transplantation shows that most if not all extrapancreatic insulin-producing cells derive from donor bone-marrow.

## STEM CELLS IN LIVER AND PANCREAS

In isolated pancreatic tissue, pancreas-resident progenitor cells might give rise to endocrine islet cells. Human and rodent pancreatic duct cells, islet derived cells and exocrine tissue contain cells that can differentiate towards a pancreatic endocrine phenotype. These tissues cultured and differentiated *in vitro*, have been transplanted and can reverse diabetes mellitus in rodents. Rodent liver stem cells and human fetal-liver cells have been differentiated *in vitro* into insulin secreting cells by culture methods and or introduction of cell specific genes. When transplanted these cells reverse diabetes mellitus in rodents. Cells within liver that can differentiate into insulin secreting cells after introduction of B cell specific genes have also been seen *in vivo* after adenoviral gene delivery into rodents that have been rescued from diabetes for long periods. A bona fide pancreatic stem cell for cell regeneration remains elusive. A genetic marking study in mice casts doubt on the existence of any cell progenitor cells and pancreatic progenitor cell.

Curative therapy for diabetes mellitus mainly implies replacement of functional insulin-producing pancreatic cells, with pancreas or islet cell transplants. However, shortage of donor organs spurs research into alternative means of generating cells from islet expansion, encapsulated islet xenografts, human islet cell lines, and stem cells. Stem cell therapy here implies the replacement of disease or lost cells from progeny of pluripotent or multipotent cells. Both embryonic stem cells (derived from the inner cell mass of a blastocyst) and adult stem cells (found in the postnatal organism) have been used to generate surrogate cells or otherwise restore cell functioning. Recently, Andreas Lechner and colleagues failed to see transdifferentiation into pancreatic cells after transplantation of bone marrow cells into mice. Last year, Jayaraj Rajagopal and colleagues failed to derive cells from embryonic stem cells. As in every emerging field in biology, early reports seem confusing and conflicting. Embryonic and adult stem cells are potential sources for cell replacement and merit further scientific investigation. Discrepancies between different results need to be reconciled. Fundamental process in determining the differentiation pathways of stem cell remains to be elucidated, so that rigorous and reliable differentiation protocols can be established. Encouraging studies in rodent models may ultimately set the stage for large animal studies and translational investigation.

## ARE STEM CELLS THE NEXT FRONTIER FOR DIABETES TREATMENT?

Diabetes is a devastating disease that affects millions of people worldwide. The major forms of the disease are type 1 and type 2 diabetes. In type 1 diabetes, the body's immune system aberrantly destroys the insulin-producing beta cells (β-cells) of the pancreas. Type 2 diabetes, the more common form, is characterized both by insulin resistance, a condition in which various tissues in the body no longer respond properly to insulin action, and by subsequent progressive decline in β-cell function to the point that the cells can no longer produce enough additional insulin to overcome the insulin resistance. Researchers are actively exploring cell replacement therapy as a potential strategy to treat type 1 diabetes, because patients with this disease have lost all or nearly all β-cells were developed, such a treatment strategy could also be useful for the larger population with type 2 diabetes. One of the major challenges of cell replacement therapy is the current insufficient supply of β-cells from human organ donors. This article focuses on stem cells as potential sources for deriving new β-cells.

## DIABETES: A CRITICAL HEALTH ISSUE FOR THE 21ST CENTURY.

According to the International Diabetes Federation, diabetes currently affects 7% of the world's population—nearly 250 million individuals worldwide. This total is expected to rise to 380 million by 2025 as a result of aging populations, changing lifestyles, and a recent worldwide increase in obesity. Although projection for increase in diabetes prevalence suggests that the greatest percentage gain will occur in Asia and South America, all nations will experience a rising disease burden. According to the National Diabetes Fact sheet, which was compiled using information from the Centers for Diseases Control and Prevention and other Federal and non-Federal organizations, 20.8 million U.S. children and adults have diabetes (6.2 million of whom are currently undiagnosed). An estimated 54 million Americans have "pre-diabetes", a condition defined by blood glucose levels that are above normal but not sufficiently high to be diagnosed as diabetes. In 2005, 1.5 million new cases were diagnosed as diabetes. In 2005, 1.5

million new cases of diabetes were diagnosed in Americans aged 20 years or older. If present trends continue, 1 in 3 Americans (1 in 2 minorities) born in 2000 will develop diabetes in their lifetime.

Diabetes is currently the sixth leading cause of death in the U.S. for it is associated with numerous health complications, including increased risk for heart disease, stroke, kidney disease, blindness, and amputations. In 2007, the total annual economic cost of diabetes was estimated to be $174 billion dollars. Direct medical expenditures account for the vast majority of this total ($116 billion), although lost productivity and other indirect cost approached nearly $58 billion. The American Diabetes Association estimates that one out of every 10 health care dollars currently spent in the U.S. is used for diabetes and its complication. While diabetes can be managed, at present it cannot be cured. As a result, it is a lifelong and often disabling disease that can severely impact the quality of life of those who are afflicted. Based on several recent discoveries, however, researchers have begun to ask if a new treatment approach is on the horizon—can stem cells that are derived from adult or embryonic tissues generate new pancreatic β-cells to replace those that have failed or been destroyed? Cell replacement therapy is one of many research avenues being pursued as a potential treatment strategy for type 1 diabetes. The strategy may also have implications for ameliorating type 2 diabetes. One of the key obstacles to advancing such therapy is the current inadequate supply of cadaveric donor pancreas as a source of cells for transplantation. Additionally, it is not currently possible to induce a patient's own cells to regenerate new β-cells within the body. Thus, researches are actively investigating potential sources of new beta cells, including different types of stem cells. The article will focus on the various types of stem cells that are candidates for use in pancreatic regeneration and will discuss the challenges of using such cells as therapy for diabetes.

## Defining Diabetes

Diabetes results from the body's inability to regulate the concentration of sugar (glucose) in the blood. Blood glucose concentration is modulated by insulin, a hormone produced by pancreatic β-cells and released into the blood stream to maintain homeostasis. In healthy individuals, β-cells counteract sharp increases in blood glucose, such as those caused by a meal, by releasing in initial "spike" of insulin within a few minutes of the glucose challenge. This acute release is then followed by a more sustained release that may last for several hours, depending on the persistence of the elevated blood glucose concentration. The insulin release gradually tapers as the body's steady-state glucose concentration is re-established. While postprandial insulin release is stimulated by factors other than blood glucose, the blood sugar concentration is the major driver. When the β-cells fail to produce enough insulin to meet regulatory needs, however, the blood glucose concentration rises. This elevated concentration imposes a metabolic burden on numerous body systems, dramatically increasing the risk of premature cardiovascular disease, stroke and kidney failure. Moreover, the risk for certain diabetes related complications increase even at blood glucose concentrations below the threshold for diagnosing diabetes.

At present there is no cure for diabetes. β-cell failure is progressive once the condition is manifest, full function usually cannot be restored. Those with type 1 diabetes require daily insulin administration to survive. Persons with type 2 diabetes must control their elevated blood glucose levels through various means, including diet and exercise, oral antihyperglycemic (blood glucose-lowering) drugs, and/or daily insulin shots. Most people who live with type 2 diabetes for a period of time will eventually require insulin to survive.

As noted earlier, there are different forms of diabetes. Type 1 diabetes results when a person's immune system mistakenly attacks and destroys the β-cells. This type of diabetes was once referred to as juvenile onset diabetes, because it usually begins in childhood. Type 1 diabetes accounts for 5–10% of diabetes cases, and people with type 1 diabetes depend on daily insulin administration to survive. By contrast, type 2 diabetes is a metabolic disorder that results from a decline in β-cell function combined with insulin resistance, or the inability to use insulin effectively in peripheral tissues such as the liver, muscles, and fat. Diabetes onset is associated with genetic factors and with obesity, and type 2 diabetes disproportionately affects certain minority groups. Unlike type 1 diabetes, type 2 is largely preventable. Numerous studies have suggested that the environmental and behavioural factors that promote obesity (e.g., a sedentary lifestyle, a high calorie diet) have profoundly influenced the recent rise in the prevalence of type 2 diabetes. This trend suggests that type 2 diabetes will continue to be a major health care issue.

## THE CASE FOR STEM CELLS

There is great interest in developing strategies to expand the population of functional β-cells. Possible ways to achieve this include physically replacing the β-cell mass via transplantation, increasing β-cell replication, decreasing β-cell death, and deriving new β-cells from appropriate progenitor cells. In 1990, physicians at the Washington University Medical Center in St. Louis reported the first successful transplant of donor supplied pancreatic islet tissue (which includes β-cells) in humans with type 1 diabetes. By the end of the decade, many other transplants had been reported using various protocols, including the widely known Edmonton protocol (named for the islet transplantation researchers at the University of Alberta in Edmonton). This protocol involves isolating islets from the cadaveric pancreatic tissue of multiple donors and infusing them into the recipient's portal vein. However, the lack of available appropriate donor tissue and the strenuous regimen of immunosuppressive drugs necessary to keep the body from rejecting the transplanted tissue limit the widespread use of this approach. Moreover, the isolation process for islets damages the transplantable tissue; as such 2–3 donors are required to obtain the minimal β-cell mass sufficient for transplantation into a single recipient. While these strategies continue to be improved, islet function declines relatively rapidly post-transplant. For example, a long term follow-up study of Edmonton transplant patients indicated that less than 10% of recipients remained insulin independent five years after transplant. These challenges have led researchers to explore the use of stem cells as a possible therapeutic option. Type 1 diabetes is an appropriate candidate disease for stem cell therapy, as the causative damage is localized to a particular cell type. In theory, stem cells that can differentiate into β-cells in response to molecular signals in the local pancreatic environment could be introduced into the body, where they would migrate to the damaged tissue and differentiate as necessary to maintain the appropriate β-cell mass. Alternately, methods could be developed to coax stem cells grown in the laboratory to differentiate into insulin-producing β-cells. Once isolated from other cells, these differentiated cells could be transplanted into a patient. As such, stem cell therapy would directly benefit persons with

type-1 diabetes by replenishing β-cells that are destroyed by autoimmune processes, although it would still be necessary to mitigate the autoimmune destruction of β-cells. The strategy would also benefit those with type 2 diabetes to a lesser extent by replacing failing β-cells, although the insulin resistance in peripheral tissues would remain. As discussed in the following sections, however, debate continues about potential sources of pancreatic stem cells.

The pancreas is located in the abdomen, adjacent to the duodenum (the first portion of the small intestine). A cross-section of the pancreas shows the islet of Langerhans which is the functional unit of the endocrine pancreas. Encircled is the beta cell that synthesizes and secretes insulin. Beta cells are located adjacent to blood vessels and can easily respond to changes in blood glucose concentration by adjusting insulin production. Insulin facilitates uptake of glucose, the main fuel source, into cells from which glucose in turn enter tissues such as muscle.

## Searching for the "Pancreatic Stem Cell"

The pancreas is a complex organ made up of many cell types. The majority of its mass is comprised of exocrine tissue, which contains acinar cells that secrete pancreatic enzymes into the intestine to aid in food digestion. Dispersed throughout this tissue are thousands of islets of Langerhans, clusters of endocrine cells that produce and secrete hormones into the blood to maintain homeostasis. The insulin producing β-cell is one type of endocrine cell in the islet; other types include alpha cells (α-cells), which produce glucagon, gamma cells (γ-cells), which produce pancreatic polypeptide, and delta cells (δ-cells), which produce somatostatin. Each of these cell types arises from a precursor cell type during the process of development. Therefore, the key step for using stem cells to treat diabetes is to identify the precursor cell(s) that ultimately give rise to the β-cell. However, generating these cells is more complex than simply isolating a hypothetical pancreatic stem cell. Experiments have indicated that embryonic and adult stem cells can serve as sources of insulin-secreting cells, 16 leading researchers to explore several avenues through which stem cells could feasibly be used to regenerate β-cells. However, many challenges must be addressed before a particular cell type will become established for this approach.

The human body has inherent mechanisms to repair damaged tissue, and these mechanisms remain active throughout life. Thus, there is reason to speculate that the adult pancreas may be aided by some type of regenerative system that replaces worn-out cells and repairs damaged tissue in response to injury. Such a system could theoretically be supported by precursor or stem cells, located in the endocrine pancreas or elsewhere, which could be coaxed to differentiate in response to select molecular or chemical stimuli. But do these cells exist? If so, how can they be recognized, isolated, and cultured for therapeutic use? How quickly could they produce sufficient numbers of β-cells to offset damage caused by diabetes processes? Alternately, what if cells that have the capability to regenerate β-cells exist in the body but are committed to differentiate into some other cell type? Could embryonic stem (ES) cell lines, which have the potential to develop into cells from all lineages, be explored in the following sections, which review the types of candidate stem cells for diabetes.

## Are Adult Pancreatic Stem Cells Present in the Pancreas?

Whether β-cell progenitors are present in the adult pancreas is a controversial topic in diabetes research. Several recent studies in rodents have indicated that the adult pancreas contains some type of endocrine progenitor cells that can differentiate toward β-cells. However, researchers have not reached consensus about the origin of the bona fide pancreatic stem cell (if it exists) or the mechanism(s) by which β-cells are regenerated. For example, a pivotal study by Dor and colleagues used genetic lineage tracing in adult mice to determine how stem cells contribute to the development of β-cells. Their analysis indicated that new β-cells arise from pre-existing ones, rather than from pluripotent stem cells, in adult mice. As such, the authors noted that β-cells can proliferate *in vivo*, thereby cast(ing) doubt on the idea that adult stem cells have a significant role in β-cell replenishment. Soon after this report was published, Seaberg and coworkers reported the identification of multipotent precursor cells from the adult mouse pancreas. These novel cells proliferated *in vitro* to form colonies that could differentiate into pancreatic α-, β-, and δ-cells as well as exocrine cells, neurons, and glial cells. Moreover, the beta-like cells demonstrated glucose-dependent insulin release, suggesting possible therapeutic application to diabetes. Several subsequent studies have also reported the existence of pancreatic stem/precursor cells *in vitro* or *in vivo*. One recent report suggests that such cells exists in the pancreatic ductal lining and can be activated autonomously in response to injury, increasing the β-cell mass through differentiation and proliferation. The study of pancreatic regeneration continues to evolve, and many claims have been made regarding cells believed to be involved in the process. In the last decade, reports have described various putative pancreatic stem cells embedded in the pancreatic islets, pancreatic ducts, among the exocrine acinar cells, and in unspecified pancreatic locales in rodent models, as well as from human adult pancreatic cell lines, islet tissue, and non-islet tissues discarded after islets have been removed for transplantation. These cells are identified by the presence of one or more cell-surface proteins, or markers, known to be associated with a particular stem cell lineage. However, these studies illustrate several challenges shared by all researchers who seek to identify the "pancreatic stem cell". First, all potential stem cell candidates identified to date are relatively rare; or instance, the precursor cells identified by Seaberg are present at the rate of 1 cell per 3,000–9,000 pancreas cells. Because there are so few of these putative stem cells, they can be difficult to identify. Additionally, the choice of marker can select for certain stem cell populations while possibly excluding others. Interestingly, the progenitor cells identified in the Seaberg study lacked some known β-cells markers such as HNF3b, yet they were able to generate β-cells. Thus, a hypothetical experiment that used only HNF3b as a marker for β-cell differentiation would likely not identify this stem cell population. Moreover, techniques used to study the pancreatic tissue, such as the genetic lineage technique of Dor, et.al could possibly interfere with the generation of new β-cells from stem or precursor cells.

As such, the possibility remains that β-cells could be regenerated by differentiation of endogenous stem cells, by proliferation of existing β-cells, or a combination of the two mechanisms.

Further research to elucidate conditions under which β-cells can proliferate may help to develop new therapeutic approaches. For example, several advances have recently been made from studies of pregnancy and pregnancy-related diabetes (gestational diabetes) in mice. During pregnancy, pancreatic islet cells normally expand in number to meet increased metabolic demands. Researchers have found that the protein HNF4-alpha helps increase β-cell mass, and that pregnancy-related decrease in levels of another protein, menin, also enables β-cell proliferation. Insights may also arise from research on another organ, the liver. Unlike the pancreas, the liver has an inherently high capacity for regeneration. New strategies for including pancreatic islet cell growth may emerge from knowledge of how liver cells develop from progenitor cells during early development such that the resulting adult organ retains substantial regenerative capacity. In another research avenue, scientists are exploring whether it may be possible to redirect adult pancreatic cells in the body to change from their original cell type into β-cells.

## Other Potential Sources of Stem Cells Derived from Adult Cells

Furthermore, various reports have also described putative stem cells in the liver, spleen, central nervous system, and bone marrow that can differentiate into insulin-producing cells. While it is possible that such pathways may exist, these results are currently under debate within the research community. In another research avenue, scientists recently reported that differentiated cells, including adult human skin cells, can be genetically "reprogrammed" to revert to a pluripotent state, resembling that of embryonic stem (ES) cells. The researchers refer to these cells as induced pluripotent stem (iPS) cells. This approach may facilitate the establishment of human iPS cells lines from patients with specific diseases that could be used as research tools. The technique, or variations of it, may also one day allow patient specific stem cells to be generated for use in stem cell-based therapies. However, the genes used for reprogramming were introduced into the cells using a virus-based method, which could have adverse clinical effects. If, however, safe alternate methods based on this research can be developed for reprogramming cells, then iPS cells may lead to novel, personalized therapies.

## Can Embryonic Stem Cells Be Used?

The challenges associated with identifying and isolating adult "pancreatic stem cells" has led some researchers to explore the use of ES cells as a source of insulin-producing cells. Several factors make ES cells attractive for this application. First, given the complexity of pancreatic tissue, identified β-cell precursors would likely be difficult to isolate from the adult pancreas. If isolated, the cells would then need to be replicated *ex vivo* while keeping them directed toward a β-cell lineage. Second, protocols to grow and expand mature β-cells in culture have met with technical challenges. ES cells, which are pluripotent cell lines (they can give rise to all cell types of the embryo) that can be induced to develop into various lineages based on culture conditions, may therefore represent a future option for β-cell regeneration.

To date, several human ES cells lines have been successfully derived. While these cell lines serve as resources for exploring the mechanisms of development, their potential use in a clinical setting is limited by several factors, most notably ethical concerns and the risk of teratoma development. Researchers are only beginning to unlock the myriad factors that come into play as a once pluripotent cell differentiates into a unipotent cell, one that can contribute to only one mature cell type. For example, several recent reports indicate that mouse and human cells can be successfully differentiated into endodermal cells, the precursors of pancreatic cells. In addition, insulin-producing cells have been derived from mouse and human ES cells. However, it should be noted that directed differentiation of ES cells toward the β-cell has not been reported. Beta cells appear relatively late during embryonic development, suggesting that their presence involves the temporal control of a considerable number of genes. Moreover, the creation of patient-specific, stem cell-derived b-cells for transplantation requires genetic matching to lessen the immune response. Generating immune-matched tissues requires the therapeutic cloning of human ES cells, which has not been accomplished to date. A fraudulent claim to the contrary in 2005 by South Korean researcher Woo Suk Hwang ignited international controversy within the scientific community and illustrated the scientific and ethical challenges of using ES cells as a source of transplant tissue. Despite current gaps in knowledge, researchers recognize the potential of ES cells as sources of specialized cells such as the β-cell s, and the study of ES cells provides insight into the processes that govern differentiation and specialization.

***Preparation of embryonic stem cells*** All procedure should be carried out using sterile techniques. The growth and maintenance media for ES cells is M-15:DMEM (no pyruvate, high glucose), 15% FBS, 1X GPS, 1X BME. Handling ES cells; growth, maintenance, passing, freezing and thawing is conducted in a manner to protect and maintain the quality of the cells and keep them in a pluripotential state. Serum quality is critical for successful growth of ES cells and especially true for blastocysts. The quality of the feeder is very instrumental. Remember also that in passing, freezing and electroporating ES cells, it is best that the cells are still at exponential growth (80% confluence) for optimal results.

### *Thawing (quick thaw)*

1. Remove cells from the freezer and quickly thaw in a 37$^0$C water bath.
2. Transfer the cell suspension to a sterile 15 ml tube. Add 10–12 ml of M15 media to 1 ml of cell suspension.
3. Gently mix and pellet the cells by centrifuging at 1000 rpm for 7 minutes.
4. Aspirate off supernatant and resuspend cells into 6 ml of M15, and plate out cells in a 6-cm feeder plate.
5. Refeed cells daily with fresh M15, upon 80–85% confluence, cells need to be passaged or frozen.

### *Passaging ES cells*

ES cells typically should be passaged every 2–4 days (apart from colonies under selection). If passaging is neglected, the cells with differentiate and you will select for variants that might have lost totipotency. Cells must be fed when media begins to turn orange.

1. Check cells under the microscope for 80–85% confluence.
2. Refeed cells 3–4 hours before passing them (VERY IMPORTANT).
3. Aspirate media off. Wash one time with PBS. Add trypsin (see table below).
4. Incubate at 37$^0$C for 15 minutes.

5. Inactivate the trypsin by adding at least an equal volume of media (the serum inactivates the trypsin).

6. Pipette up and down several times to separate the cells and break any colonies.

7. Add fresh media to fresh feeder (see table below). Split ratios for ES cells can vary from 1:1 to 1:10. It is not recommended to dilute them more than 1:10.

The area relationships for the various dishes are as follows:

| Dish | Media | Trypsin | Area ($cm^2$) | Diameter in cm (actual) |
|---|---|---|---|---|
| 96 well plate | 200 µl/well | 30–50µl | 0.3 | 0.6 |
| 24 well plate | 1 ml/well | 200 µl | 1.8 | 1.5 |
| 6 well plate | 3–4 ml | 400 µl | 9.6 | 3.5 |
| 6 cm dish | 6ml | 0.6 ml | 21.2 | 5.2 |
| 10 cm dish | 12 ml | 1.5 ml | 60 | 8.7 |
| 15 cm dish | 30 ml | 3 ml | 154 | 14 |

Some typical passaging ratios:

1:6=1 x 60 mm to 2 x 90 mm

1:6=1 x 30 mm to 1 x 90 mm

1:4=1 x 30mm to 2 x 60mm

1:5=1 x 24 mm to 1 x 30 mm (6 well plate)

1:6 = 1 x 96 well to 1 x 24 well.

8. Aliquot the cell suspension into plates in the volume specified for each plate. Always check the feeder before using them. They should be confluent, no gaps, not contaminated and not dividing. Use feeders that are older (1–2 weeks old), the advantages are many; any contamination is assessed, also any dividing run-away cells can be detected, and the passage will be earlier. Also, older feeder would have settled nicely and flattened.

9. Mix carefully, moving the plate in a cross (up and back, then sideways) to distribute the cells evenly. Swirling them is not a good idea.

## Freezing ES Cells (Slow Freeze)

1. Check cells under the microscope. 80–85% confluence is best.
2. Refeed cells 2–4 hours before freezing them.
3. Aspirate media off. Wash one time with PBS. Add an appropriate amount of trypsin.
4. Incubate at $37^0$ C for 15 minutes
5. Inactivate the trypsin by adding at least an equal volume of media (the serum inactivates the trypsin).
6. Pipette up and down several times to separate the cells and break any colonies. Collect cell suspension in a centrifuge tube and note the volume.
7. Count a 200 µl aliquot and calculate the total cell number. From this calculate the volume of media required to give a final density of $3.0 \times 10^7$ cells/ml
8. Pellet cells by centrifuging at 1000 rpm for 7 minutes.
9. Aspirate off supernatant and resuspend the pellet in half the volume calculated in Step 7 of M15.
10. Add half the volume with 2x Freezing Media (60% DMEM, 20% FBS, 20% DMSO, freshly prepared); the cell suspension is diluted as a result 10% DMSO is the final concentration. Add the freezing media dropwise, mixing well after each addition.
11. Aliquot the suspension into sterile freezing vials, pre-labelled with cell type (AB2.2, AB1, etc.,) clone number, passage number and date. A typical aliquot would have 0.3 ml–0.4 ml of ES cells. This is about a total of $9 \times 10^6$–$12 \times 10^6$ cells total/vial.

12. Place vials into a freezing container. Optimally, the temperature should drop about 1°C per minute, and the isopropanol freezing container are best for this application. Freeze cells overnight at 80°C.

13. Next day, transfer cells to the Liquid Nitrogen freezer.

## CLINICAL CHALLENGES

Clearly using stem cells to treat diabetes will require additional knowledge, both in the laboratory and in the clinic. This section will suggest several envisioned approaches for stem-cell derived diabetes therapies and discuss key considerations that must be addressed for their successful application.

Contingent upon the development of appropriate protocols, stem cells could theoretically be used to treat diabetes through two approaches. Both strategies would require the isolation and *in vitro* expansion of a homogeneous population of β-cell precursor cells from appropriate donor tissue. Once a population of these cells has been generated, they could either, 1) be induced to differentiate into insulin-producing cells *in vitro* and then be transplanted into the diabetic patient's liver, or 2) be injected into the circulation along with stem cell stimulators, with the hope that the cells will "home in" to the injured islets and differentiate into a permanent self-renewing β-cell population.

Because type 1 diabetes is an autoimmune disease, controlling the autoimmune response is critical to the success of any potential stem cell-based therapy. Type 1 diabetes is characterized by the action of β-cell specific, auto reactive T cells. Even if the regenerative properties of the pancreas remain functional, the continued presence of these T-cells effectively counteracts any endogenous repair and would likely decimate populations of newly regenerated or transplanted insulin-producing cells. However, the autoimmune response has been successfully averted in non-obese diabetic mice either by using anti-T-cell antibodies to eliminate the majority of the autoreactive cells or by transplanting bone marrow from a diabetes-resistant donor (with a sublethal dose of irradiation) into the diabetic animal. Both strategies appear to enable the replenishment of insulin-secreting cells and the eventual restoration of normal blood glucose levels, although the process requires weeks to months and may necessitate additional therapy. Other strategies being explored include altering the immune tolerance through the use of monoclonal antibodies, proteins, and oligonucleotides. Other clinical challenges, including safety, tracking of stem cells, delivery of the cells to the targeted tissue within a clinically relevant time frame (for transplanted cells), identification of ways to promote long term survival and functioning of regenerated β-cell s, ease of obtaining the cells and cost, parallel those encountered with all applications of stem cell based regenerative therapy . These issues must be addressed once the "pancreatic stem cell" population has been identified conclusively. Given current debate on this issue, the routine clinical application of stem cell based regenerative therapy for the treatment of diabetes remains a future goal, albeit one with great potential.

As an additional source of information, an extensive discussion of research challenges and strategies for achieving the goal of cell replacement therapy for type 1 diabetes is presented in Advance and Emerging opportunities in Type 1 Diabetes Research.

The results discussed in this article demonstrate the many challenges that must be addressed before stem cells can be used to regenerate islet tissue in persons with diabetes. Debate continues on the identification of the pancreatic stem cell and at present it is difficult to ascertain which cell type has the greatest potential for diabetes therapy. Moreover, modulating the auto immune response in type 1 diabetes remains a significant challenge regardless of the type of cell that is transplanted, and it will also be important to address the insulin resistance in type 2 diabetes, as well as factors that contribute to obesity. However, diabetes is a disease with a major deficiency in the functioning of one type of cell, and there is potential of stem cells to treat type 1 diabetes and to improve the quality of life for those with type 2 diabetes. As researchers learn more about the mechanisms that govern stem cell programming, differentiation and renewal, their ability to identify, isolate, and culture candidate stem cells will continue to improve. While stem cells can be currently considered a frontier for diabetes therapy, they may one day become its basis.

# 21

# Gene Therapy

## USE OF GENETICALLY MODIFIED STEM CELLS IN EXPERIMENTAL GENE THERAPIES

Gene therapy is a novel therapeutic branch of modern medicine. Its emergence is a direct consequence of the revolution heralded by the introduction of recombinant DNA methodology in 1970s. Gene therapy is still highly experimental, but has the potential to become an important treatment regimen. In principle, it allows the transfer of genctic information into patient's tissues and organs. Consequently, diseased genes can be eliminated or their normal functions rescued. Furthermore, the procedure allows the addition of new functions to cells, such as the production of immune system mediator proteins that help to combat cancer and other diseases.

Originally, monogenic inherited diseases (those caused by inherited single gene defects) such as cystic fibrosis were considered primary targets for gene therapy.

While the positive therapeutic outcome was celebrated as a breakthrough for gene therapy, a serious drawback subsequently became evident. By February 2005, three children out of seventeen who had been successfully treated for X-linked SCID developed leukaemia because the vector inserted near an oncogene (a cancer causing gene) inadvertently caused it to be inappropriately expressed in the genetically-engineered lymphocyte target cell. On a more positive note, a small number of patients with adenosine deaminase-deficient SCID have been successfully treated by gene therapy without any adverse side effects.

A small number of more recent gene therapy clinical trials, however, are concerned with monogenic disorders. The majority of current clinical trials (66% of all trials ) focus on polygenic diseases, particularly cancer.

Gene therapy relies on similar principles as traditional pharmacologic therapy, specifically, regional specificity for the targeted tissues, specificity of the introduced gene function in relation to disease and stability and controllability of expression of the introduced gene. To integrate all these aspects into a successful therapy is an exceedingly complex process that requires expertise from many disciplines, including molecular and cell biology, genetics and virology, in addition to bio-process manufacturing capability and clinical laboratory infrastructure.

## THE TWO PATHS TO GENE THERAPY

Gene therapy can be performed either by direct transfer of genes into the patient or by using living cells as vehicles to transport the genes of interest. Both modes have certain advantages and disadvantages. Direct gene transfer is particularly attractive because of its relative simplicity. In this scenario, genes are delivered directly into a patient's tissues or blood stream by packaging into liposomes (spherical vessels composed of the molecules that form the membrane of cells) or other biological microparticles. Alternately, the genes are packaged into genetically-engineered viruses, such as retroviruses or adenoviruses. Because of bio-safety concerns, the viruses are typically altered so that they are not toxic or infectious, that is, they are replication-incompetent. These basic tools of gene therapy have been extensively optimized over the past 10 years.

However their biggest strength—simplicity is simultaneously their biggest weakness. In mean cases, direct gene either randomly integrates into the patient's chromosomes or persists unintegrated for a relatively short period of time in the targeted tissue. Additionally, the targeted organ or tissue is not always easily accessible for direct application of the therapeutic gene.

On the other hand, therapeutic genes can be delivered using living cells. The procedure is relatively complex in comparison to direct gene transfer, and can be divided into three major steps. In the first step, cells from the patient or other sources are isolated and propagated in the laboratory. Second, the therapeutic gene is introduced into these cells, applying methods similar to those used in direct gene transfer. Finally, the genetically-modified cells are returned to the patient. The use of cells as gene transfer vehicles has certain advantages. In the laboratory dish (*in vitro*), cells can be manipulated much more precisely than in the body (*in vivo*). Some of the cell types that continue to divide under laboratory condition may be expanded significantly before reintroduction into the patient. Moreover, some cell types are able to localize to particular regions of the human body, such as hematopoietic (blood forming ) stem cells, which return to the bone marrow. This quot; homingquot phenomenon may be useful for applying the therapeutic gene with regional specificity.

A major disadvantage, however, is the additional biological complexity brought into systems by living cells. Isolation of a specific cell type requires not only extensive knowledge of biological markers but also insight into the requirements for that cell type to stay alive *in vitro* and continue to divide. Unfortunately, specific biological markers are not known for many cell types, and the majority of normal human cells cannot be maintained for long periods of time *in vitro* without acquiring deleterious mutations.

## STEM CELLS AS VEHICLES FOR GENE THERAPY

Stem cells can be classified as embryonic or adult, depending on their tissue of origin. The role of adult stem cells is to sustain an established repertoire of mature cell types in an essentially steady-state number over the life time of the organism. Although adult tissues with a high turnover rate, such as blood, skin, and intestinal epithelium, are maintained by tissue-specific stem cells, the stem cells themselves rarely divide. However, in certain situations, such as during tissue repair after injury or following transplantation, stem cell divisions, may become more frequent. The prototypic example of adult stem cells, the hematopoietic stem cell has already been demonstrated to be of utility in gene therapy. Although they are relatively rare in the human body, these cells can be readily isolated from bone marrow or after mobilization into peripheral blood. Specific surface markers allow the identification and enrichment of hematopoitic cells from a mixed population of bone marrow or peripheral blood cells.

After *in vitro* manipulation, these cells may be retransplanted into patients by injection into the blood stream, where they travel automatically to the place in the bone marrow in which they are functionally active. Hematopoietic stem cells that have been explanted, manipulated *in vitro* and re-transplanted into the same patient (autologous transplantation) or a different patient (allogeneic transplantation) retain the ability to contribute to all mature blood cell types of the recipient for an extended period of time. (When patients cell are temporarily grown quot; outside the body quot; before being returned to them, the in vitro process is typically referred to as a quot; ex vivoquot; approach). Another adult bone marrow-derived stem cell type with potential use as a vehicle for gene transfer is the mesenchymal stem cell, which has the ability to form cartilage, bone adipose (fat) tissue, and marrow stroma (the bone marrow microenvironment). Recently, a related stem cell type, the multipotent adult progenitor cell, has been isolated from bone marrow that can differentiate into multiple lineages, including neurons, hepatocytes (liver cells), endothelial cells (such as the cells that form the lining of blood vessels ), and other cell types. Other adult stem cells have been identified such as those in the central nervous system and heart, but these are less well characterized and not as easily accessible.

The traditional method to introduce a therapeutic gene into hematopoietic stem cells from bone marrow or peripheral blood involves the use of a vector derived from a certain class of virus, called a retrovirus. One type of retroviral vector was initially employed to show proof-of-principle that a foreign gene (in that instance the gene was not therapeutic, but was used as a molecular tag to genetically mark the cells) introduced into bone marrow cells may be stably maintained for several months. However, these particular retroviral vectors were only capable of transferring the therapeutic gene into actively dividing cells. Since most adult stem cells divide at a relatively slow rate, efficiency was rather low. Vectors derived from other types of retroviruses (lentiviruses) and adenoviruses have the potential to overcome this limitation, since they also target non-dividing cells.

The major drawback of these methods is that the therapeutic gene frequently integrates more or less randomly into the chromosomes of the target cell. In principle, this is dangerous because the gene therapy vector can potentially modify the activity of neighbouring genes by integrating into them. These phenomena are referred to as quot insertional mutagenesis quot; in extreme cases such as in the X-linked SCID gene therapy trials, these mutations contribute to the malignant transformation of the targeted cells, ultimately resulting in cancer.

Another major limitation of using adult stem cell is that it is relatively difficult to maintain the stem cell state during *ex vivo* manipulations. Under current suboptimal conditions, adult stem cells tend to lose their stem cell properties and become more specialized, giving rise to mature cell types through a process termed quot; differentiation. Quot. Recent advances in supportive culture conditions for mouse hematopoietic stem cells may ultimately facilitate more effective use of human hematopoietic stem cells in gene therapy applications.

## Embryonic stem cells: "The Ultimate Stem Cell"

Embryonic stem cells are capable of unlimited self-renewal while maintaining the potential to differentiate into derivatives of all three germ layers. Even after months and years of growth in the laboratory, they retain the ability to form any cell type in the body. These properties reflect their origin from cells of the early embryo at a stage during which the cellular machinery is geared toward the rapid expansion and diversification of cell types.

Murine (mouse) embryonic stem cells were isolated over 20 years ago and paved the way for the isolation of non-human primate, and finally human embryonic stem cells. Much of the anticipated potential surrounding human embryonic stem cells is an extrapolation from pioneering experiments in the mouse system. Experiments performed with human embryonic stem cells in the last couple of years indicates that these cells have the potential to make an important impact on medical science, at least in certain fields. In particular, this impact includes,

1) differentiation of human embryonic stem cells into various cell types such as neurons, cardiac, vascular, hematopoietic, pancreatic, hepatic, and placental cells,

2) the derivation of new cell lines, under alternative conditions and,

3) the establishment of protocols that allow the genetic modification of these cells.

### *The potential of human embryonic stem cells for gene therapy*

Following derivation, human embryonic stem cells are easily accessible for controlled and specific genetic manipulation. When this facility is combined with their rapid growth, remarkable stability, and ability to mature *in vitro* into multiple cell types of the body, human embryonic stem cells are attractive potential tools for gene therapy. Two possible scenarios whereby human embryonic stem cells may benefit the gene therapy field are discussed below.

First, human embryonic stem cells could be genetically manipulated to introduce the therapeutic gene. This gene may either be active or awaiting later activation once the modified embryonic stem cell has differentiated into the desired cell type. Recently published reports establish the feasibility of such an approach where skin cells from an immunodeficient mouse were used to generate cellular therapy that partially restored immune function in the mouse. In these experiments, embryonic stem cells were generated from an immundeficient mouse by nuclear transfer technology. The nucleus of an egg cell was replaced with that from a skin cell of an adult mouse with the genetic immunodeficiency. The egg was developed to the blastula stage at which embryonic stem cells were derived. The genetic defect was corrected by a genetic modification strategy designated quot; gene targeting quot; these quot; cured quot; embryonic stem cell were differentiated into hematopoietic quot; stemquot; cells and transplanted into immune deficient mice. Interestingly the immune function in these animals was partially restored. In principle, this approach may be employed for treating human patients with immunodeficiency or other diseases that may be corrected by cell transplantation.

However, significant advances must first be made. The levels of immune system reconstitution observed in the mice were quite modest (<1% of normal) while the methodology employed to achieve hematopoietic engraftment is not clinically feasible. This methodology involved using a more severely immunodeficient mouse as recipient (which also had the murine equivalent of the human X-linked SCID mutation) and genetically engineering the hematopoietic engrafting cells with a potential oncogene prior to transplantation.

Embryonic stem cells may additionally be indirectly beneficial for cellular gene therapy. Since these cells can be differentiated *in vitro* into many cell types, including presumably tissue-specific stem cells, they may provide a constant *in vitro* source of cellular material. Such quot; adult Quot stem cells derived from embryonic stem cells may thus be utilized to optimize protocols for propagation and genetic manipulation techniques. To acquire optimal cellular materials from clinical samples in larger quantities for experimental and optimization purposes is usually rather difficult since access to these samples is limited.

## Genetic Manipulation of Stem Cells

The therapeutic gene needs to be introduced into the cell type used for therapy. Genes may be introduced into cells by transfection or transduction. Transfection utilizes chemical or physical methods to introduce new genes into cells. Usually small molecules, such as liposomes, as well as other cationic to lipid-based particles are employed to facilitate the entry of DNA encoding the gene of interest into the cells. Brief electric shocks are additionally used to facilitate DNA entry into living cells. All of these techniques have been applied to various stem cells, including human embryonic stem cells. However, the destiny of the introduced DNA is relatively poorly controlled using these procedures. In most cells, the DNA disappears after days or weeks and in rare cases, integrates into host chromosomal DNA. *In vitro* drug strategies allow the isolation and expansion of cells that are stably transfected as long as they significantly express the newly introduced gene. Transduction utilizes viral vectors, for DNA transfer. Viruses, by nature introduce DNA or RNA into cells very efficiently. Engineered viruses can be used to introduce almost any genetic information into cells. However, there are usually limitations in the size of the introduced gene. Additionally some viruses (particularly retroviruses) only infect dividing cells effectively, where as other (lentiviruses) do not require actively dividing cells. In most cases the genetic information carried by the viral vector is stably integrated into the host cell genome (the total complement of chromosomes in the cell). An important parameter that must be carefully monitored is the random integration into the host genome, since this process can induce mutations that lead to malignant transformation or serious gene dysfunction. However several copies of the therapeutic gene may also be integrated into the genome, helping to bypass positional effects and gene silencing. Positional effects are caused by certain areas within the genome and directly influence the activity of the introduced gene. Gene silencing refers to the phenomenon whereby over time, most artificially introduced active genes are turned off by the host cell, a mechanism that is not currently well understood. In these cases, integration of several copies may help to achieve stable gene expression, since a subset of the introduced genes may integrate into favourable sites. In the past, gene silencing and positional effects were a particular problem in mouse hematopoietic stem cells. These problems led to the optimization of retroviral and lentiviral vector systems by the addition of genetic control elements (referred to as chromatin domain insulators and scaffold/matrix attachment regions) into the vectors, resulting in more robust expression in differentiating cell systems, including human embryonic stem cells. In some gene transfer systems, the foreign transgene does not integrate at a high rate and remains separate from the host genomic DNA, a status denoted quot; episomalquot; Specific proteins stabilizing these episomal DNA molecules have been identified as well as viruses (adenovirus) that persist stably for some time in an episomal condition. Recently, episomal systems have been applied to embryonic stem cells.

An elegant way to circumvent positional effects and gene silencing is to introduce the gene of interest specifically into a defined region of the genome by the gene targeting technique. The gene targeting technique takes advantage of a cellular DNA repair process known as homologous recombination. Homologous recombination provides a precise mechanism for defined modifications of genomes in living cells, and has been used extensively with mouse embryonic stem cells to investigate gene function and create mouse model of human diseases. Recombinant DNA is altered *in vitro*, and the therapeutic gene is introduced into a copy of the genomic DNA that is targeted during this process. Next, recombinant DNA is introduced by transfection into the cell, where it recombines with the homologous part of the cell genome. This, in turn results in the replacement of normal genomic DNA with recombinant DNA containing genetic modification.

Homologous recombination is a very rare event in cells, and thus a powerful selection strategy is necessary to identify the cells in which it occurs. Usually, the introduced construct has an additional gene coding for antibiotic resistance (referred to as a selectable marker) allowing cells that have incorporated the recombinant DNA to be positively selected (i.e. to keep only those thymocytes which have a T cell receptor capable of binding MHC (Major Histo Compatibility)) in culture. However, antibiotic resistance only reveals that the cells have taken up recombinant DNA and incorporated it somewhere in the genome. To select for cells in which homologous recombination has occurred, the end of the recombination construct often includes the thymidine kinase gene from the herpes simplex virus. To select for cells in which homologous recombination has occurred, the end of the recombination construct often includes the thymidine kinase gene from the herpes simplex virus. Cells that randomly incorporate recombinant DNA usually retain the entire DNA construct, including the herpes virus thymidine kinase gene. In cells that display homologous recombination between the recombinant construct and cellular DNA, an exchange of homologous DNA sequences is involved, and the non-homologous thymidine kinase gene at the end of the construct is eliminated. Cells expressing the thymidine kinase gene are killed by the antiviral drug ganciclovir in a process known as negative selection (i.e. the elimination of autoreactive thymocytes). Therefore these cells undergoing homologous recombination are unique in that they are resistant to both the antibiotic and ganciclovir allowing effective selection with these drugs.

Gene targeting by homologous recombination has recently been applied to human embryonic stem cells. This is important for studying gene functions *in vitro* for lineage selection and marking. For therapeutic applications in transplantation medicine, the controlled modification of specific genes should be useful for purifying specific embryonic stem cell derived, differentiated cell types from a mixed population, altering the antigenicity of embryonic stem cell derivatives, and adding defined markers that allow the identification of transplanted cells. Additionally, since the therapeutic gene can now be introduced into defined regions of the human genome, better controlled expression of the therapeutic gene should be possible. This also significantly reduces the risk of insertional mutagenesis.

## FUTURE CHALLENGES FOR STEM CELL-BASED GENE THERAPY

Despite promising scientific results with genetically modified stem cells, some major problems remain to be overcome. The more specific and extensive the genetic modification, the longer the stem cells have to remain *in vitro*. Although human embryonic stem cells in the culture dish remain remarkably stable, the cells may accumulate genetic and epigenetic changes that might harm the patient (epigenetic changes regulate gene activity without altering the genetic blueprint of the cell – which is harmful). Indeed sporadic chromosomal abnormalities in human embryonic stem cell culture have been reported, and these may occur more frequently when the cells are passaged as bulk populations. This observation reinforces the necessity to optimize culture conditions further, to explore new human embryonic stem cell lines, and to monitor the existing cell lines. Additionally undifferentiated embryonic stem cells have the potential to form a type of cancer called a teratocarcinoma. Safety precautions are therefore necessary and currently, protocols are being developed to allow the complete depletion of any remaining undifferentiated embryonic stem cells. This may be achieved by rigorous purification of embryonic stem cell derivatives or introducing suicide genes that can be externally controlled.

Another issue is the patient's immune system response. Transgenic genes, as well as vectors introducing these genes (such as those derived from viruses) potentially trigger immune system responses. If stem cells are not autologous, they eventually cause immuno-rejection of the transplanted cell type (due to mismatching of the cells). Strategies to circumvent these problems such as the expression of immune system modulating genes by stem cells, creation of chimeric, immunotolerable bone marrow or suppression of HLA genes have been suggested in this context. Nuclear transfer technology has been recently extended to human embryonic stem cell, notably, immune-matched human embryonic stem cells have now been established from patients, including an individual with an immunodeficiency disease, congenital hypogammaglobulinemia. Strategies that combine gene targeting with embryonic stem cell-based therapy are thus potential novel therapeutic options.

The addition of human embryonic stem cells to the experimental gene therapy arsenal offers great promise in overcoming many of the existing problems of cellular based gene therapy that have been encountered in clinical trials. Further research is essential to determine the full potential of both adult and embryonic stem cells in this exciting new field.

# 22

# Stem cell Treatment for Crohn's Disease

Cellular therapy with stem cells is revolutionizing the focus of treatment of many serious diseases. Replacing the cells of damaged tissue with other new cells from the same patient is already a reality. This is the basis of cellular therapy and regenerative medicine, the latest great advance in biomedicine. In this line, Hospital Clinic, Barcelona is leading the world in the application of an innovative cellular therapy that uses stem cells to treat Crohn's disease, a chronic genetic disease that affects 1% of the population in Spain and which has considerable impact on the quality of life of the patients. The procedure is based on an autologous bone-marrow transplant (when patients receive a transplant of their own stem cells) and now constitutes a treatment option to cure an intestinal disease that sometimes does not successfully respond to drugs and requires highly complex surgery that does not provide a cure.

## CELLULAR THERAPY AS A STRATEGY TO COMBAT CROHN'S DISEASE

Crohn's disease, together with ulcerative colitis, is included in what is called irritable bowel disease. It is a chronic genetic disease that occurs when the immune system loses tolerance to the patient's own intestinal flora, leading to an abnormal inflammatory response that continues over time. The results are inflammation and ulceration in different areas of the digestive tract, leading to the symptoms. The disease progresses in the form of unpredictable and variable out break throughout the patient's life and the severity of the symptoms varies according to the level of involvement of the intestines and the patient's response to the assigned treatment. It is a disease that usually affects young people between the ages of 18 and 40 years. Diagnosis is often difficult because it presents symptoms similar to those of other diseases of the digestive tract—bdominal pain, diarrhoea, vomiting, nausea, fever, general malaise, etc.

Hospital Clinic de Barcelona is one of the few hospitals in the world to instate cellular therapy using autologous stem cell transplantation. In the US, the treatment has been tested on 12 patients with severe Crohn's disease, of whom 11 have had very good results; in Italy the treatment has been applied to 4 patients, 3 of whom are also showing excellent progress following the transplant.

In severe cases involving recurrent outbreaks (reactivation of the disease several times throughout the patient's life) Crohn's disease presents several treatment options. Firstly, doctors choose to use corticosteroids and immunosuppressant and biological drugs to control the inflammatory process and prevent complications of the disease such as stenosis ( narrowing of the intestinal lumen) or fistulas (openings from the intestinal lumen to other organs, such as the intestine, bladder, vagina or skin). However, over the course of the disease as much as 70% of severe patients require surgery to remove segments affected by the disease, due to failure of the pharmacologic approach. The surgery is occasionally very aggressive for the patient as it is sometimes necessary to remove the entire colon or large sections of the small intestine, thereby considerably affecting the absorption of food by the intestine, with a resulting deterioration in quality of life and body image (colostomy bag). For this reason new treatments are being developed for patients in whom the solution has not been an option to date.

## Autologous Stem Cell Transplant: Phases of the Procedure

When the case is detected (that does not respond to drugs or surgery), the patient undergoes an autologous stem cell transplant, which is a bone marrow transplant in which the immune system is reset to prevent it from attacking the intestinal flora. The process lasts approximately 2 months and consists of 6 phases.

1. **Initial chemotherapy (cyclophosphamide + G-CSF)** In this initial phase, leucopenia or reduction of the number of leukocytes (immune system cells) in the blood is induced in the patient.

2. **Migration of stem cells to the blood** Following the previous immune suppression, the organism reacts by releasing stem cells from the bone marrow into the blood. These are the cells which will later be used for the transplant.

3. **Collection of stem cells by means of apheresis** Apheresis is a technique that separates components of the blood. Here, the stem cells that previously migrated from the bone marrow are separated.

4. **Cryopreservation of stem cells** When the stem cells have been collected by apheresis, they are frozen and preserved until ready for transplant.

5. **Second chemotherapy** In this phase, total leucopenia is induced, that is, the immune system is left devoid of leukocytes, ready to be reset with the stem cell transplant.

6. **Autologous stem cell transplant** The patient receives the transplant by means of transfusion with his or her own stem cells. The immune system is reset, leading to remission or reduction of the abnormal inflammatory process of Crohn's disease.

# 23

# Neuronal Stem Cells Tracked Using MRI

The recently patented technology could be used to further the study of neural stem cells and inform the development of new treatments for brain injury caused by trauma, stroke, Parkinson's disease and other neurological disorders.

Legend had it that once a brain cell dies, it's lost forever. Neuroscientists now know that this is purely myth, having proved that the brain is constantly producing new neurons. These neural stem cells are born deep in an area of the brain called the subventricular zone. As time goes on, the cells also called neuroblasts, make their way to other areas of the brain where they mature into functioning neurons. The brain's ability to regenerate its cells is of great interest to scientists. "If we could better understand the molecular migratory signals that guide neuroblasts, we could try to redirect these cells to areas of the brain harmed by stroke or traumatic brain injury. With this information scientists might be able to one day repair the brain", said Ahrens, who also is a member of the Pittsburgh NMR Centre for Biomedical Research.

Studying cells in a living brain is problematic. Common forms of *in vivo* cell imaging like fluorescene and bioluminescence rely on light to produce images, making them unsuitable for viewing neuroblasts buried deep beneath the skull and layers of opaque tissue. Until now scientists had only been able to study neuronal stem cells by looking at slices of the brain under a microscope. Ahrens was able to surmount this problem using MRI technology. Rather than light, MRI uses magnets to create high resolution images. A typical MRI scan uses a magnetic field and radio frequency pulses to cause the hydrogen protons found in the body's water molecules to give off signals. Those signals are converted into a high resolution image.

As reported in a 2005 issue of Nature Medicine, Ahrens developed a method that causes cells to produce their own contrast agent allowing them to be imaged with MRI. Using a viral vector, Ahrens incorporated the genes that produce the naturally occurring metalloprotein ferritin into living cells. Ferritin which is present in all biological cells, harvests and stores naturally occurring iron. When the cells tagged with ferritin began to produce increased amounts of the protein, they draw in additional iron, turning themselves into nanomagnets. This disrupts the magnetic field surrounding the tagged cells, changing the signal given off by adjacent water molecules. This change appears as dark spots on the MRI image indicating the presence of cells. Since then, Ahrens team has improved on the process, developing an engineered form of ferritin that is a more effective MRI reporter than naturally occurring ferritin.

In the current study, Lordanova and Ahrens used the same technique as in the initial study, this time tagging neuroblasts with the engineered ferritin. They incorporated the DNA sequence for the engineered metalloprotein into an adenovirus vector, which they then injected into the subventricular zone of a rat brain. The adenovirus infected the neural stem cells giving the cells the genetic instructions to being producing the ferritin reporter. Lordanova then imaged the brain with MRI and found that she was able to follow—in real time—the neuroblasts as they travelled towards the olfactory bulb and ultimately formed new inhibitory neurons. These results mirrored what had been observed in histology studies. Recently, scientists of Carnegie mellon University received a patent for the reporter. Ahrens hopes to continue to develop the technology in order to allow researchers to better understand neuronal stem cells and how neurons regenerate. Ahrens also plans to use the reporters to improve clinical trials of cell-based therapies. By incorporating the MRI reporter into the cells before implantation, researchers would be able to find the answer to a number of critical questions as given below.

Where do these cells go, days, weeks and months later? How do we know that they have grafted to the right cells? Or have they grafted in the wrong place? Or died? All these questions need to be answered in future.

## VIRUS-FREE EMBRYONIC LIKE STEM CELLS MADE FROM SKIN OF PARKINSON'S DISEASE PATIENTS.

Science Daily (Mar 8, 2009), Whitehead institute researchers have developed a novel method to remove potential cancer-causing genes during the reprogramming of skin cells from Parkinson's disease patients into an embryonic stem cell like state. Scientists then used the resulting induced pluripotent stem (iPS) cells to derive dopamine-producing neurons, the cell type that degenerates in Parkinson's disease patients.

This marks the first time researchers have generated human iPS cells that have maintained their embryonic stem cell like properties after the removal of reprogramming genes. The findings are published in the March 6 edition of the journal Cell.

"Until this point, it was not completely clear that when you take out the reprogramming genes from human cells, the reprogrammed cells would actually maintain the iPS state and be self-perpetuating", says Frank Soldner, a postdoctoral researcher in Whitehead Member Rudolf Jaenishch's laboratory and co-author of the article.

Since August 2006, researchers have been reprogramming adult cells into iPS cells by using viruses to transfer four genes (Oct4, Sox2, c-Myc and Klf4) into the cell's DNA. Although necessary for reprogramming cells, these genes, the known oncogene c-Myc in particular, also have the potential to cause cancer. In addition, the four genes interact with approximately 3000 other genes in the cell, which may change how the cell functions.

In the current method, Whitehead researchers used viruses to transfer the four reprogramming genes and a gene coding for the enzyme Cre into skin cells from Parkinson's disease patients. The reprogramming genes were bracketed by short DNA sequences called *loxP*, which are recognized by the enzyme Cre. After the skin cells were reprogrammed to iPS cells the researchers introduced the Cre enzyme into the cells, which removed the DNA between the two *loxP* sites, thereby deleting the reprogramming genes from the cells. The result is a collection of iPS cells with genomes virtually identifical to those of the Parkinson's disease patients from whom original skin cells came.

Removing the reprogramming genes is also important because of those genes' effect on an iPS cell's gene expression (a measure of which genes the cell is using and how much it's using those genes). When the researchers compared the gene expressions of human embryonic stem cells to iPS cells with and without the reprogramming factors, iPS cells without the reprogramming genes had a gene expression closer to human embryonic stem cells than to the same iPS cells that still contained the reprogramming genes.

"The reprogramming factors are known to bind to and affect the expression of 3,000 genes in the entire genome, so having artificial expression of those genes will change the cell's overall gene expression", Dirk Hockemeyer, who is also a co-author of the Cell article. That is why the four reprogramming genes can mess up the system so much. From now on, it will be tough for researchers to leave the reprogramming genes in iPS cells.

Jaenisch says that the process to remove the reprogramming genes is very successful, when compared with earlier experiments. After removing the reprogramming genes, the Jaenisch researchers differentiated the cells from the Parkinson's disease patients into dopamine-producing nerve cells. In Parkinson's disease patients, these cells in the brain die or become impaired, causing such classic Parkinson's symptoms as tremors, slowed movement and balance problems. Because the cells reside in the patient's brains researchers cannot easily access them to investigate how the disease progresses at the cellular level, what kills the cells, or what might prevent cellular damage. Therefore the ability to create patient-specific iPS cells, derive the dopamine-producing cells, and study those patient-specific cells in the laboratory could be a great advantage for Parkinson's disease researchers. Although the initial results are extremely promising, Jaenisch acknowledges that the process is far from over. The next step is to use these iPS-derived cells as disease models.

Whitehead institute researchers reprogrammed human skin cells from Parkinson's disease patients into an embryonic stem cell-like state. After removing the reprogramming genes, the scientists used these so called induced pluripotent stem (iPS) cells to create dopamine-producing neurons, the cell type that degenerates in Parkinson's disease patients. To confirm that the iPS cells had become dopamine-producing neurons, the researchers stained the cells green for a neuron-specific protein (class III betatubulin) and red for a dopamine-producing neuron specific enzyme (tyrosine hydroxylase);

## RELATED REFERENCE

- Stem cells Reverse Disease in a model of Parkinson's disease (May 16, 2011)—Researchers have compared the ability of cells derived from different types of human stem cell to reverse disease in a rat model of Parkinson disease and identified a stem cell population.
- Skin Cells could Help Discover Cause of Parkinson's Disease (July 16, 2010 )—Researchers are applying new stem cell technology to use skin samples to grow the brain cells thought to be responsible for the onset of Parkinson's disease.
- New Technique Produces Genetically Identical stem cells (July 6, 2008)—Cells from mice created using genetically reprogrammed cells can be triggered via drug administration to enter an embryonic stem cell-like state.
- Reprogrammed Human Blood Cells show promise for Disease Research (July 5, 2010)—Cells from frozen human blood samples can be reprogrammed to an embryonic stem cell-like state according to new research.

# 24

# Congenital Neurology Disorders

Neurological congenital malformations are not rare. Many infants are born with abnormalities like neural tube defect, hydrocephalus, microcephaly and spinal defect. There is high mortality and morbidity due to neurological birth defects.

The common congenital neurological diseases are neural tube defects like spina bifida, meningocele, myelomeningocele, encephalocele, etc. The other common diseases are hypoxic ischemic encephalopathy secondary to prenatal asphyxia, neonatal seizures, intraventricular haemorrhage especially in preterms, congenital viral and bacterial infections resulting in meningitis. The rare ones are neurological manifestations secondary to metabolic and degenerative diseases, neurocutaneous syndromes and epilepsy syndromes.

Further, providing statistical information on congenital neurological anomalies, congenital neurological diseases come with high morbidity and mortality rates. Perinatal asphyxia has 15–45 per cent morbidity and upto 30 per cent mortality. Neural tube defects carry a mortality risk of 15 per cent in developed countries and the risk is high in developing countries such as India. It is very difficult to access prevalence due to lack of reporting in our country. However, the prevalence of perinatal asphyxia is around 2 per cent of live births in the US. Neural tube defects have a prevalence of 1 in 2000 live births and intraventricular haemorrhage occurs in 5 per cent of newborns who are less than 35 weeks of gestation.

Overall the incidence of these anomalies is coming down due to the better diagnostic facilities, prenatal care and good maternal nutrition. Because of early diagnosis and appropriate treatment, the mortality has significantly reduced. However, the morbidity still persists at a lower level. The incidence is 1–3 per 1000 live births. Incidence varies depending on the geographical regional factors and consanguinity.

## CAUSATIVE FACTORS

A large number of congenital neurological problems are possible in the brain and spinal cord. Many of these are related to genetic abnormalities, while the others are related to formative problems. "Genetics plays a very important role in the occurrence of these diseases. Some genetically inherited diseases include disease of the brain, muscles and nerves."

"Neurocutaneous syndromes, such as neurofibromatosis, tuberous sclerosis and Sturge-Weber disease are familial. Sporadic gene mutations cause disorders of neuronal migration, such as lissencephaly, schizencephaly and holoprosencephaly. Various syndromes with central nervous system (CNS) involvement have a genetic abnormality such as Aicardi Syndrome, hydrocephalus secondary to aqueductal stenosis and genetic epileptic syndromes."

Other factors responsible for these congenital neurological disorders include the prenatal condition of the mother, drug intake during pregnancy, consanguinity, birth injuries and pre-existing congenital problems in the family. "Extreme prematurity can cause intraventricular haemorrhage and hydrocephalus. Perinatal asphyxia resulting from hypoxic ischaemic encephalopathy, and mental retardation with developmental delay are other common pathologies found in India. Various viral infections like toxoplasmosis, cytomegalovirus ans herpes simplex as well as bacterial infections from *Escherichia coli*, *Klebsiella* and pneumococcus are still predominant in our country resulting in meningitis seizures."

## DIAGNOSIS, PROGNOSIS AND MANAGEMENT

An early diagnosis of congenital neurological disorders can be done, especially for neural tube defects. Prenatal screening, including an ultrasound and certain blood tests will help in diagnosing these diseases, giving a prognosis to the family and deciding on further line of treatment after birth, in terms of surgery.

Now-a-days it is routine to have prenatal ultrasound scan around the 15$^{th}$ week of gestation to identify anomalies. If there is a suspicion, then amniotic fluid analysis, genetic counselling, metabolic screening and foetal MRI scan are various modalities for diagnosis. Majority of the abnormalities can be diagnosed except some genetic and metabolic syndromes.

"Computer-based investigations have remarkably changed the diagnostic scenario in neurological problems. Computed Tomography (CT) scan and Magnetic Resonance Imaging (MRI) have revolutionized the investigations in the last two decades. Foetal scanning can also be informative regarding gross problems in the foetus. Evaluation and understanding of these problems can sometimes lead to attempts towards premature termination of pregnancy or surgery *in utero*."

Generally, the prognosis in neurological diseases depends on the type of anomaly, how early it is diagnosed and how well it is treated. In addition, rehabilitation of the child, motivation of parents and family, and development of complications will also influence the prognosis. As far as the prognosis is concerned, obviously the outcome is bad in metabolic and degenerative neurological diseases as there is still no cure for majority of them. With the availability of specific anticonvulsants, outcome is better in epileptic case. The improvement in neonatal surgeries and post-operative care of surgical case such as neural tube defects has certainly improved the growth and development of these patients, but the morbidity and mortality remains high.

Neurological diseases, if diagnosed timely and treated appropriately, the long term outcome in many of these congenital diseases is satisfying. However, there are numerous congenital anomalies, each having treatment specific to it. There is no uniform treatment for congenital disorders in neurology. Surgical procedures like shunt surgery, fixation surgery and repair surgery have helped patients. Other procedures like foetal surgical treatment is yet not popular in India, such a treatment is not done.

## ROLE OF STEM CELLS

Stem cells are widely believed to have significant potential in the treatment of human diseases. They also offer new opportunities for the treatment of incurable neurological diseases. Emphasizing on the role of stem cells in congenital malformation, Dr. Arvin Faundo, group medical affairs mananger, Cordlife group, says, "Microcephaly is a congenital malformation of the nervous system, and there is no specific treatment for this condition, except supportive care. Some patients with microcephaly have been associated with Fanconi anaemia. The latter is the result of a defect in the ability of a person for deoxyribonucleic acid (DNA) repair. It is manifested by bone marrow failure and eventual development of cancers. Definitive treatment involves bone marrow transplantation with a suitable donor. After the first successful cord blood transplant for Fanconi anaemia in 1988, efforts are now placed on improving outcomes for these patients."

"A relatively common neurologic condition that may benefit from stem cell therapy is cerebral palsy. This refers to disorders manifesting with developmental disabilities in both movements. It may be related to damage to the motor centres of the infantile brain before, during or after childbirth." Elaborating further, Dr. Faundo adds, "The extrapolated incidence in India is about two million children. Management of the most severe cases requires life-long supportive therapy. Following reports of physical and psychosocial improvements in patients, who received stem cell infusions, several clinical trials are underway to evaluate the benefit of autologous cord blood cells."

Dr. Venkataramana feels, "Stem cells seem to be promising in promoting regeneration, thereby repairing the damaged areas. This will definitely reduce the disability. Stem cells have immunomodulatory qualities and may be beneficial in modifying disease profile. However, a lot more needs to be understood regarding the application of stem cell therapy to these congenital diseases affecting the brain and spinal cord."

Dr. Venkataramana also says that "Several trials in the world are conducted to prevent the anomaly by nutritional supplements, for example, fortification of folic acid to prevent spina bifida. However, for early treatment, several trails are conducted to treat the anomaly in order to prevent further implications and complications. It is now a separate specialty called foetal neurosurgery."

Further, he adds, "Further challenges will be to identify all the causative factors for these anomalies and understand the genetic factors in order to prevent the complications by altering these factors. Gene therapy and regenerative medicine like stem cells can play a significant role in future in reducing the disability caused by congenital neurological disorders."

## STEM CELL TRAIL TO COMBAT CHILDHOOD BRAIN DISEASE

A team from the Oregon Health and Science University and Doernbecher Children's Hospital plan to treat six children suffering from the inherited neurodegenerative condition, Batten's diseases—also known as neuronal ceriod lipofuscinosis (NCL). There is currently no alternative treatment for the disease.

To start with, the children receive injections of neural stem cells that have been purified–isolated from other cell types—and grown from donated human fetal tissue.

Children with Batten's disease suffer seizures, motor control disturbances, blindness with communication problems. As many as 600 children in the US are currently diagnosed with the condition—death can occur in children as young as 8 years old.

These children lack an enzyme for breaking down complex fat and protein compounds in the brain, explains Robert Steiner, Vice-chair of paediatric research at the hospital. The material accumulates and interferes with tissue function, ultimately causing brain cells to die.

## Neuron Support

Previous tests on animals demonstrated that stem cells injected into the brain secreted the missing enzyme. And the stem cells were found to survive well in the rodent brain.

Once injected, the purified neural cells may develop into neurons or other nervous system tissue, including oligodendrocytes, or glial cells, which support the neurons. Steiner is hopeful that the treatment will work for the 25 or so other hereditary brain diseases related to Batten's diseases.

In addition to secreting enzymes, Steiner says these cells can become the type of nerve cells found in spinal cord, and so they could potentially help after spinal cord injury. The stem cells can form into neural cells found in the brain or nerve cells found elsewhere in the central nervous system.

The researchers hope that the treatment will offer some clinical benefit to the children suffering from Batten's disease, but they stress that the primary purpose of the trail is to assess the safety of the product to be approved by the FDA. Human fetal stem cell transplants have been performed before on adult patients with neurodegenerative conditions including Parkinson's and Huntington's diseases, and also for spinal cord injury.

The researchers plan to follow the children's progress over the course of a year.

## NEUROGENESIS

Neurogenesis (birth of neurons) is the process by which neurons are generated from neural stem and progenitor cells. Most active during pre-natal development neurogenesis is responsible for populating the growing brain with neurons.

New neurons are continually born throughout adulthood in predominantly two regions of the brain:

- The **sub ventricular zone** (SVZ) lining the lateral ventricles, where neural stem cells and progenitor generate new neurons (neuroblast) that migrate to the olfactory bulb via to rostral migratory stream.

- The **sub granular** zone (SGZ), part of the dentate gyrus of hippocampus

Many of the newborn cells die shortly after they are born, but a number of them become functionally integrated into the surrounding brain tissue.

Adult neurogenesis is an example of a long-held scientific theory being overturned. Early neuroanatomists, including Santiago Ramon Y Cajal, considered the nervous system fixed and incapable of regeneration. The first evidence of adult mammalian neurogenesis in the cerebral cortex was presented by Joseph Altman in 1962. In 1963 the adult neurogenesis was demonstrated in the dentate gyrus of the hippocampus. Shirley Bayer also showed that adult neurogenesis exists in mammals (rats), and Nottebohm showed the same phenomenon in birds. More recently neurogenesis in the cerebellum of adult rabbits has also been characterized. Further, some authors have suggested that adult neurogenesis may also occur in regions with the brain not generally associated with neurogenesis including the neocortex. However others have questioned the scientific evidence of these findings, arguing that the new cells may be of glial origin.

## Role of Neurogenesis in Learning

The functional relevance of adult neurogenesis is uncertain, but there is some evidence that hippocampal adult neurogenesis is important for learning and memory. Multiple mechanisms for the relationship between increased neurogenesis and improved cognition have been suggested including computational theories to demonstrate that new neurons increase memory capacity.

Studies have demonstrated that the act of learning itself is associated with increased neuronal survival. However the overall findings that adult neurogenesis is important for any kind of learning are equivocal.

## Effects of Stress

Adult-born neurons appear to have a role in regulation of stress. Studies have linked neurogenesis to the beneficial actions of specific antidepressants, suggesting a connection between decreased hippocampal neurogenesis and depression. In a subsequent paper scientists demonstrated that the behavioural benefits of antidepressant administration in mice is reversed when neurogenesis is prevented with x-irradiation techniques. In fact, new born neurons are more excitable than older neurons due to a differential expression of GABA receptors. These neurons augment the role of the hippocampus in the negative feedback mechanism of the HPA-axis (physiological stress) and perhaps in inhibiting the amygdala (the region of brain responsible for fearful responses to stimuli). Indeed, suppression of adult neurogenesis can lead to an increased HPA-axis stress response in mildly stressful situations. This is consistent with numerous findings linking stress-relieving activities (learning, exposure to a new yet benign environment, and exercise) to increased levels of neurogenesis. Moreover the animals exposed to physiological stress (cortisol) or psychological stress (e.g. isolation) show markedly decreased levels of new born neurons. This elevation of newborn neurons by antidepressants improves, under chronic stress conditions, the hippocampal control on the stress response (including the activity of the HPA axis and of stress-integrative brain nuclei), then leading to recovery without newborn neurons.

Some studies have hypothesized that learning and memory are linked to depression, and that neurogenesis may promote neuroplasticity. One study proposes that mood may be regulated, at a base level, by plasticity, and thus not chemistry. Accordingly, the effects of antidepressant treatment would only be secondary to change in plasticity.

## Effects of Sleep Reduction

One study has linked lack of sleep to a reduction in rodent hippocampal neurogenesis. The proposed mechanism for the observed decrease was increased level of glucocorticoids. It was shown that two weeks of sleep deprivation acted as a neurogenesis-inhibitor, which was reversed after return of normal sleep and even shifted to a temporary increase in normal cell proliferation.

## Possible use in Treating Parkinson's Disease

Recent studies show the studies evidence for the existence of neural stem cells with the potential to produce new neurons, particularly of a dopaminergic phenotype, in the adult mammalian brain. Proliferation is restored completely by a selective agonist of D2 like (D2L) receptors. Neural stem cells have been identified not only in the neurogenic brain regions, where neurogenesis is constitutively ongoing, but also in the non-neurogenic zones, such as the midbrain and the striatum, where neurogenesis does not occur under normal physiological conditions. These studies lead to elegant cell therapies for neurodegenerative disorders such as Parkinson's disease by mobilizing autologus endogenous neural stem cells to replace degenerated neurons.

## Role in Behavioural Sensitization

Reinforcing drugs such as amphetamines and opiates induce behavioural sensitization upon repeated administration by inducing dopaminergic neurogenesis in the ventral tegmental area (VTA) and substantia nigra pars compacta (SNC) and this occurs through activation of dopamine receptors in these areas which produces glutamate release and subsequent elevation of local basic fibroblast growth factor (bFGF) concentrations. The consequences of these actions provide a potential response to stabilize the behavioural sensitization

## Effects of Exercise

Scientists have shown that physical activity in the form of voluntary exercise results in an increase in the number of newborn neurons in the hippocampus of aging mice. Though the association between exercise-mediated neurogenesis and enhancement of learning remains unclear, this study clearly demonstrates the benefits of physical activity and could have strong implications in the fields of aging and/or Alzheimer's disease.

## Changes in Old Age and Alzheimer's Disease

Allopregnanolone, a neurosteroid, aids the continued neurogenesis in the brain. Levels of allopregnanolone in the brain decline in old age and Alzheimer's disease. Allopregnanolone has been shown through reversing impairment of neurogenesis to reverse the cognitive deficits in a mouse model of Alzheimer's diseases.

## REGULATION OF NEUROGENESIS

Many factors may affect the rate of hippocampal neurogenesis. Exercise and an enriched environment have been shown to promote the survival of neurons and the successful integration of newborn cells into the existing hippocampus. Another factor is central nervous system injury since neurogenesis occurs after cerebral ischemia, epileptic seizures, and bacterial meningitis. On the other hand, conditions such as chronic stress and aging can result in a decreased neuronal proliferation.

## ADULT NEURAL STEM CELLS

Neural stem cells (NSCs) are the self-renewing, multipontent cells that generate the main phenotypes of the nervous system.

## EFFECTS OF MEDICINAL MARIJUANA

Some students have shown that use of cannabis results in the growth of new nerve cells in the hippocampus from both embryonic and adult stem cells. In 2005, a clinical study of rats at the University of Saskatchewan showed regeneration of nerve cells in the hippocampus.

# 25

# Stem Cells and Cardiac Repair

## MESENCHYMAL (BONE MARROW STROMAL) CELLS

Mesenchymalstemcells (MSCs) are precursors of non-haematopoietic tissues (e.g.,muscle, bone, tendons, ligaments, adipose tissue, and fibroblasts) that are obtained relatively easily from autologous bone marrow. They remain multipotent following expansion *in vitro*, exhibit relatively low immunogenicity, and can be frozen easily. While these properties make the cells amenable to preparation and delivery protocols, scientists can also culture them under special conditions to differentiate them into cells that resemble cardiac myocytes. This property enables their application in cardiac regeneration. MSCs differentiate into endothelial cells when cultured with vascular endothelial growth factor and cardiomyogenic (CMG) cells when treated with the DNA-demethylating agent, 5-azacytidine. More important, however is the observation that MSCs can differentiate into cardiomyocytes and endothelial cells *in vivo* when transplanted to the heart following myocardial infarct (MI) or non-injury in pig, mouse, or rat models. Additionally, the ability of MSCs to restore functionally may be enhanced by the simultaneous transplantation of other stem cell types. Several animal model studies have shown that treatment with MSCs significantly increases myocardial function and capillary formation. One advantage of using these cells in human studies is their low immunogenicity; allergenic MSCs injected into infarcted myocardium in a pig model regenerated myocardium and reduced infarct size without evidence of rejection. A randomized clinical trial implanting MSCs after MI has demonstrated significant improvement in global and regional LV function and clinical trials are currently underway to investigate the application of allogeneic and autologous MSCs for acute MI and myocardial ischemia, respectively.

## RESIDENT CARDIAC STEM CELLS

Recent evidence suggests that the heart contains a small population of endogenous stem cells that most likely facilitates minor repair and turnover mediated cell replacement. These cells have been isolated and characterized in mouse, rat and human tissues. The cells can be harvested in limited quantity from human endomyocardial biopsy specimens and can be injected into the site of infarction to promote cardiomyocyte formation and improvements in systolic function. Separation and expansion *ex vivo* over a period of weeks are necessary to obtain sufficient quantities of these cells for experimental purposes. However, their potential as a convenient resource for autologous stem cell therapy has led the National Heart, Lung, and Blood institute to fund forthcoming clinical trials that will explore the use of cardiac stem cells for myocardial regeneration.

## ENDOTHELIAL PROGENITOR CELLS

The endothelium is a layer of specialized cells that lines the interior surface of all blood vessels (including the heart). This layer provides an interface between circulating blood and the vessel wall. Endothelial progenitor cells (EPCs) are bone marrow derived stem cells that are recruited into the peripheral blood in response to tissue ischemia. EPCs are precursor cells that express some cell surface markers characteristic of mature endothelium and some of hematopoietic cells. EPCs home in on ischemic areas, where they differentiate into new blood vessels; following a heart attack, intravenously injected EPCsproceed to the damaged region within 48 hours. The new vascularization induced by these cells prevents cardiomyocyte apoptosis (programmed cell death) and LV remodelling, thereby preserving ventricular function. However, no change has been observed in non-infarcted regions upon EPC administration. Clinical trials are currently underway to assess EPC therapy for growing new blood vessels and regenerating myocardium.

## OTHER CELLS: UMBILICAL CORD BLOOD STEM CELLS, FIBROBLASTS, AND PERIPHERAL BLOOD CD34 CELLS

Several other cell populations, including umbilical cord blood (USB ) stem cells, fibroblasts (cells that synthesize the extracellular matrix of connective tissues) and peripheral blood CD34 cells, have potential therapeutic uses for regenerating cardiac tissue. Although these cell types have not been investigated in clinical trials of heart disease, preliminary studies in animal models indicate several potential applications in humans.

Umbilical cord blood contains enriched populations of hematopoietic stem cells and mesencyhmalprecursor cells relative to the quantities present in adult blood or bone marrow when injected intravenously into the trial vein in a mouse model of MI. Human mononuclear UCB cells formed new blood vessels in the infracted heart. A human DNA assay was used to determine the migration pattern of the cells after injection; although they homed only to injured areas within the heart, they were also detected in the marrow, spleen and liver. When injected directly into the infracted area in a rat model of MI, human mononuclear UCB cells improved ventricular function.Staining for CD34 and other markers found on the cell surface of hematopoietic stem cells indicated that some of the cells survived in the myocardium. Results similar to these have been observed following the injection of human unrestricted somatic stem cells from UCB into a pig MI model.

Adult peripheral blood CD34 cells offer the advantage of being obtained relatively easily from autologous sources. Although some studies using a mouse model of MI claim that these cells can trans-differentiate into cardiomyocytes, endothelial cells and smooth muscle cells at the site of tissue injury, this conclusion is highly contested. Recent studies that involve the direct injection of blood- borne or bone marrow derived hematopoietic stem cells into the infarcted region of a mouse model of MI found no evidence of myocardial regeneration following injection of either cell type, instead, these hematopoietic stem cells followed traditional differentiation patterns into blood cells within the microenvironment of the injured heart. Whether these cells will ultimately find application in myocardial regeneration remains to be determined.

Autologous fibroblasts offer a different strategy to combat myocardial damage by replacing scar tissue with a more elastic, muscle-like tissue and inhibiting host matrix degradation. The cells may be manipulated to express muscle specific transcription factors that promote their differentiation into myotubes such as those derived from skeltonmyoblasts. One month after these cells were implanted into the post infraction scar in a rat model of MI they occupied a large portion of the scar but were not functionally integrated. Although the effects on ventricular function were not evaluated in this study, authors noted that modified autologous fibroblasts may ultimately prove useful in elderly patients who have a limited population of autologous skeletal myoblasts or bone marrow stem cells.

## CONSIDERATION FOR USING THESE STEM CELLS IN THE CLINICAL SETTING

As these examples indicate, many types of stem cells have been applied to regenerate damaged myocardium. In select applications, stem cells have demonstrated sufficient promise to warrant further exploration in large scale controlled clinical trials. However the current breadth of application of these cells has made it difficult to compare and contextualize the results generated by the various trials. Most studies published to date have enrolled fewer than 25 patients and the studies vary in terms of cell types and preparations used, methods of delivery, patient populations, and trial outcomes. However, the mixed results that have been observed in these studies donot necessarily argue against using stem cells for cardiac repair. Rather preliminary results may illuminate the gaps in understanding of the mechanisms by which these cells regenerate myocardial tissue and argue for improved characterizations of cell preparations and delivery methods to support clinical applications.

Although stem cells appear to be relatively safe in the majority of recipients to date, and increased frequency of non-sustained ventricular tachycardia, an arrhythmia, has been reported in conjunction with the use of skeletal myoblasts, future clinical trials that use stem cells for myocardial repair must address two concerns that accompany the delivery of these cells.

1) Safety, and
2) Tracking the cells to their ultimate destination(s).

## STEM CELLS PROCEDURE

Stem cell procedure starts with two programs by the combination of stem cell implantation and the self stem cells activation and proliferation process, which provides the most fertile and appropriate environment in the brain for the new implanted cells to survive and renew themselves.

Added to that, every medical program consists of a comprehensive rehabilitation program of physical and occupational therapy in our rehabilitation centre.

1. Stem cell activation and proliferation treatment

   A complete approach of treating different types of neurological malfunctioning based on cell theory, through a comprehensive treatment using medicines that stimulate the production of the body's own neural stem cells. The procedures involved are daily IV of medicines, oral medication, combined comprehensively with the rehabilitation program.

2. Stem cell implantation through different surgical procedures

   The stem cells may be injected either through the CSF (spinal cord fluid) by a lumbar puncture or through a brain injection into the damaged area, depending on the medical conditions of the patient.

## STEM CELL EXPANSION

For stem cells or even fully differentiated cell populations to be truly beneficial or useful, either therapeutically as drug screening tools, or for core research purposes, expansion of cells is a prerequisite. An appropriate culture environmentis critical to stem cell expansion efforts. This environment is typically maintained by a combination of media supplements and reagents and may or may not involve the need for health feeder population or conditioned media taken off feeder cells.

The researchers of Asian Stem Cell Regeneration Institute manipulate the autologous adipose-derived stem cells (mostly mesenchymal) extensively and change their essential character to cells that are more useful in treating your particular disease or condition (neural, cardiac or other tissue).This process is not performed in the USA or other locales.

Natural substances and biologics and/or additional devices are utilized for the best results of scienceand technology.

All operations are carried out in sterile and certified containment conditions to assure the integrity and safety of the process.

## VESCELL PERIPHERAL ARTERY DISEASE TREATMENT

Vescell PAD is the latest generation adult stem cell technology developed by Thera Vitae. VesCell PAD uses a patient's own adult stem cells to treat severe peripheral artery disease of the lower limbs and is a viable therapy for patients who have exhausted all other treatment options short of amputation.

VesCell PAD uses adult stem cells that are harvested from approximately ½ pint of the patient's own blood. In the laboratory, these adult stem cells are transformed into a therapeutic dose of cells called angiogeiccell precursors (ACPs) which are stem cells that indicate the growth of blood vessels in diseased areas of the body. Emerging research suggest that ACPs may also form additional cell types and secrete compounds conducive for growth.

The first generation of ACPs, VesCell, has been to successfully treat over 200 heart disease patients.

VesCell PAD was developed to save lower limbs that previously may have been amputated by enabling the patient's own body to repair damaged blood vessels and grow new ones, thus restoring blood flow and oxygen to damaged tissues.

Vascular surgeons are currently using VesCell PAD to improve the quality of life of patients' sufferings from severe lower limb ischemia by relieving debilitating symptoms such as claudication (leg pain) and digital gangrene.

## Treatment Procedure

All procedures are performed by medical professionals at the various hospitals affiliated with Theravitae.

Each procedure starts at the hospitals with a medical examination to confirm the applicability of treatment to the patient. Then in a process similar to donating blood, 250 ml (about ½ pint) of blood is drawn and couriered to our laboratory under strictly controlled sterile conditions.

Because there are far too few stem cells in anyone's blood to be effective, Thera Vitae developed a proprietary technology to generate millions of ACPs from the patient's blood sample. This takes about a week.

When the newly grown ACPs return from the laboratory, a vascular surgeon implants them into the affected lower limbs via intramuscular injection. Once in place, these cells now have the potential to repair damaged blood vessels and build new ones, thus restoring blood flow to the damaged tissues.

## Result of the Treatment

The treating vascular surgeon will review the medical records. This starts with the personal consultant who will assist in gathering everything that is needed. There is no charge or obligation for the entire medical review process.Normally the reviewing physician will assess the recent diagnostic results. If they are not available, he or she will request a specific test.

Vescell PAD therapy costs29,500USD for one leg and 34,500 USD for both legs.The entire procedure takes approximately 11 days and it depends on the physical condition of the patients.The treatment is quite straightforward. After the blood is drawn, it is then couriered to TheraVitae's laboratory in Israel where it is processed.

In about a week the adult stem cells will be returned from the laboratory and immediately implanted. For the next two days the patient has to remain in the hospital for observation.The treating physician will prepare a full medical report just before discharge and the patient is reviewed periodically.

## REAL TRIALS, REAL DATA, REAL IMPROVEMENT AND NO SIDE EFFECTS

This centre has successfully treated 500 cardiovascular patients with 70% success, zero deaths and side effects limited to a day or two of headaches or nausea in about 10% of the cases.

Most of these symptoms disappear within 2–3 months after VesCell stem cells are implanted. According to independent dataof over 200 patients queried, regaining their energy is the #1 benefit of stem cells, 85% of patients showed improved symptoms and quality of life and over 70 percent had objective data such as ejection fraction improvement, reduction in heart size and longer six minute walks. Average improvement is one NYHA Class on the heart chart, from Class III (Severe physical impairments) down to Class II (Moderate physical impairments).

These stem cells are very hard to prepare in sufficient numbers because they are hand-made in special (very special) laboratories only in Israel and no one else on earth can make them, so others settle for cheaper versions which work less efficiently.

## LEARN BEFORE DECISION

The patient will be contacted within 48 hours by a Patient Care Coordinator who will inform about cost, benefits, travel details and other support information. You will need to get your medical records together so a cardiologist may review them to see if you can benefit from the treatment. This is without any charge.

This procedure is an alternative therapy and is applicable in 100% of heart failure too.

This is exactly as outlined by Dr. Linus Pauling – America's leading Medical Scientist.

# Glossary

**Adult stem cells** An undifferentiated cell found in differentiated tissue that can renew itself and (with certain limitations) differentiate to yield all the specialized cell types of the tissue from which it originated.

**Blastocyst** A preimplantation embryo of about 150 cells. The blastocyst consists of sphere made up of an outer layer of cells (trophectoderm), a fluid-filled cavity (blastocoels), and a cluster of cells on the interior (inner cell mass).

**Bone marrow stromal cells (BMSC)** A stem cell found in bone marrow that generates bone, cartilage, fat, and fibrous connective tissue.

**Cell-based therapies** Treatment in which stem cells are induced to repair damaged or depleted adult cell populations or tissues.

**Clone** A line of cells that is genetically identical to the originating cell; in this case, a stem cell.

**Cumulus cell** Specialized cell that clings to ovum after ovulation that nurtures developing egg in ovary.

**Differentiation** The process whereby an unspecialized early embryonic cell acquires the features of a specialized cell such as a heart, liver, or muscle cell.

**Directed differentiation** Manipulating stem cell culture conditions to induce differentiation into a particular cell type.

**Ectoderm** Upper, outermost layer of a group of cells derived from the inner cell mass of the blastocyst that forms skin, nerves and brain.

**Early embryo (pre-embryo)** The term used to describe the preimplantation embryo that is biologically defined as the stages of development from fertilization until the appearance to the primitive streak (approximately 14 days after fertilization).

**Embryo** The developing organism from the time of fertilization until the end of the eighth week of gestation, when it becomes to be known as a foetus.

**Embryonic germ (EG) cells** Cells found in a specific part of the embryo/ foetus called the gonadal ridge that normally develop into mature gametes.

**Embryonic stem (ES) cells** Primitive (undifferentiated) cells derived from inner cell mass of a blastocyst-stage embryo that have the potential to become a wide variety of specialized cell types (pluripotent).

**Embryonic stem cell line** Embryonic stem cells, which have been cultured under *in vitro* conditions that allow proliferation without differentiation for months to years.

**Endoderm** Lower layer of a group of cells derived from the inner cell mass of the blastocyst that forms lungs and digestive organs.

**Feeder layer** Cells used in co-culture to maintain pluripotent stem cells. Cells usually consist of mouse embryonic fibroblasts.

**Fertilization** The process whereby male and female gametes unite.

**Foetus** A developing human from usually two months after conception to birth.

**Genomic imprinting** Epigenetic modifications of DNA or proteins surrounding DNA (e.g. histones) that result in parent-specific expression or repression of genes in offspring.

**Hematopoietic stem cell** A stem cell from which all red and white blood cells develop.

**Homologous recombination** Technique used to introduce exogenous genetic material into recipient nucleus based on DNA homology.

*In vitro* Literally, "in glass"; in a laboratory dish or test tube; an artificial environment.

*In vitro* **fertilization (IVF)** An assisted reproduction technique in which fertilization is accomplished outside the body.

**Inner cell mass (ICM)** The cluster of cells inside the blastocyst. These cells give rise to the embryonic disc of the later embryo and, ultimately, the foetus.

**Long-term self-renewal** The ability of stem cells to renew themselves by dividing into the same non-specialized cell type over long periods (many months to years) depending on the specific type of stem cell.

**Mesenchymal stem cells** Cells from the immature embryonic connective tissue. A number of cell types come from mesenchymal stem cells, including chondrocytes, which produce cartilage.

**Mesoderm** Middle layer of a group of cells derived from the inner cell mass of the blastocyst; it gives rise to bone, muscle, and connective tissue.

**Microenvironment** The molecules and compounds such as nutrients and growth factors in the fluid surrounding a cell in an organism or in the laboratory, which are important in determining the characteristics of the cell.

**Multipotent** Ability of a single cell to develop into many different cell types of the body. Developmental potential of a multipotent stem cell is more restricted than pluripotent or totipotent stem cell.

**Neural stem cell** A stem cell found in adult neural tissue that can give rise to neurons, astrocytes, and oligodendrocytes.

**Parthenogenesis** Form of nonsexual reproduction in which unfertilized ovum is artificially induced to develop into an embryo.

**Passage** A round of cell growth and proliferation in cell culture.

**Plasticity** The ability of stem cells from one tissue to generate the differentiated cell types of another tissue.

**Pluripotent** Ability of a single cell to develop into different cell types derived from each of the embryonic lineages (ectoderm, mesoderm, or endoderm) but cannot develop into an embryo on its own.

**Proliferation** Expansion of a population of cells by the continuous division of a single cell into two identical daughter cells.

**Reproductive cloning** A term used to describe the process of generating a human embryo via somatic cell nuclear transfer for the specific purpose of creating a cloned human being.

**Regenerative or reparative medicine** A treatment in which stem cells are induced to differentiate into the specific cell type required to repair damaged or depleted adult cell populations or tissues.

**Somatic cell nuclear transfer (SCNT)** Technique by which a somatic cell nucleus is transplanted into an ovum whose own nucleus has been removed (enucleated). Process also referred to as nuclear transfer or transplantation.

**Stromal cells** Non-blood cells derived from blood organs, such as bone marrow or fetal liver, which are capable of supporting growth of blood cells *in vitro*. Stromal cells that make this matrix within the bone marrow are also derived from mesenchymal stem cells.

**Subculturing** The process of growing and replating cells in tissue culture for many months.

**Supernumerary embryo** Embryo originally generated for the purpose of assisted reproduction by *in vitro* fertilization (INF) techniques but subsequently not used.

**Surface markers** Surface proteins that are unique to certain cell types, which are visualized using antibodies or other detection methods.

**Teratoma** A tumour composed of multiple tissues including tissues not normally found in organ in which it arises. Neoplasms frequently occur in ovary and testis. Produced experimentally in animals by injecting pluripotent stem cells, in order to determine the stem cells' abilities to differentiate into various types of tissues.

**Therapeutic cloning** A term used to describe the process of generating a human embryo via somatic cell nuclear transfer for the specific purpose of obtaining stem cells for use in regenerative medicine.

**Totipotent** Ability of a single cell to differentiate into any type of cell in the body (somatic, germ, or extraembryonic) and thus capable of forming a new organism (e.g., fertilized ovum).

**Transdifferentiation** The process by which stem cells from one tissue are able to differentiate into cells of another tissue.

**Trophoblast** The extraembryonic tissue responsible for implantation, developing into the placenta, and controlling the exchange of oxygen and metabolites between mother and embryo.

# Index

## L



## M



## N



## O



## P



## R

Made in United States
North Haven, CT
04 November 2026

21719565R00150